Provenance Data in Social Media

Synthesis Lectures on Data Mining and Knowledge Discovery

Editors
Jiawei Han, *UIUC*
Lise Getoor, *University of Maryland*
Wei Wang, *University of North Carolina, Chapel Hill*
Johannes Gehrke, *Cornell University*
Robert Grossman, *University of Chicago*

Synthesis Lectures on Data Mining and Knowledge Discovery is edited by Jiawei Han, Lise Getoor, Wei Wang, Johannes Gehrke, and Robert Grossman. The series publishes 50- to 150-page publications on topics pertaining to data mining, web mining, text mining, and knowledge discovery, including tutorials and case studies. The scope will largely follow the purview of premier computer science conferences, such as KDD. Potential topics include, but not limited to, data mining algorithms, innovative data mining applications, data mining systems, mining text, web and semi-structured data, high performance and parallel/distributed data mining, data mining standards, data mining and knowledge discovery framework and process, data mining foundations, mining data streams and sensor data, mining multi-media data, mining social networks and graph data, mining spatial and temporal data, pre-processing and post-processing in data mining, robust and scalable statistical methods, security, privacy, and adversarial data mining, visual data mining, visual analytics, and data visualization.

Privacy in Social Networks
Elena Zheleva, Evimaria Terzi, and Lise Getoor
2012

Community Detection and Mining in Social Media
Lei Tang and Huan Liu
2010

Ensemble Methods in Data Mining: Improving Accuracy Through Combining Predictions
Giovanni Seni and John F. Elder
2010

Modeling and Data Mining in Blogosphere
Nitin Agarwal and Huan Liu
2009

Provenance Data in Social Media

Geoffrey Barbier, Zhuo Feng, Pritam Gundecha, and Huan Liu

ISBN: 978-3-031-00776-7   paperback
ISBN: 978-3-031-01904-3   ebook

DOI 10.1007/978-3-031-01904-3

A Publication in the Springer series
SYNTHESIS LECTURES ON DATA MINING AND KNOWLEDGE DISCOVERY

Lecture #7
Series Editors: Jiawei Han, *UIUC*
           Lise Getoor, *University of Maryland*
           Wei Wang, *University of North Carolina, Chapel Hill*
           Johannes Gehrke, *Cornell University*
           Robert Grossman, *University of Chicago*
Series ISSN
Synthesis Lectures on Data Mining and Knowledge Discovery
Print 2151-0067   Electronic 2151-0075

Provenance Data in Social Media

Geoffrey Barbier
Air Force Research Laboratory

Zhuo Feng
Arizona State University

Pritam Gundecha
Arizona State University

Huan Liu
Arizona State University

SYNTHESIS LECTURES ON DATA MINING AND KNOWLEDGE DISCOVERY
#7

ABSTRACT

Social media shatters the barrier to communicate anytime anywhere for people of all walks of life. The publicly available, virtually free information in social media poses a new challenge to consumers who have to discern whether a piece of information published in social media is reliable. For example, it can be difficult to understand the motivations behind a statement passed from one user to another, without knowing the person who originated the message. Additionally, false information can be propagated through social media, resulting in embarrassment or irreversible damages. Provenance data associated with a social media statement can help dispel rumors, clarify opinions, and confirm facts. However, provenance data about social media statements is not readily available to users today. Currently, providing this data to users requires changing the social media infrastructure or offering subscription services. Taking advantage of social media features, research in this nascent field spearheads the search for a way to provide provenance data to social media users, thus leveraging social media itself by mining it for the provenance data. Searching for provenance data reveals an interesting problem space requiring the development and application of new metrics in order to provide meaningful provenance data to social media users. This lecture reviews the current research on information provenance, explores exciting research opportunities to address pressing needs, and shows how data mining can enable a social media user to make informed judgements about statements published in social media.

KEYWORDS

social computing, social media, provenance, data mining, social networking, microblogging

This effort is dedicated to my family. Thank you all for your support.   *– GB*

To my parents and wife.   *– ZF*

To my parents, brother, and wife.   *– PG*

To my parents, wife, and sons.   *– HL*

Contents

Acknowledgments

The authors wish to acknowledge the members of the Data Mining and Machine Learning laboratory at Arizona State University for their motivating influence and thought-inspiring comments and questions with reference to this topic. The authors also gratefully acknowledge the assistance of Suhas Ranganath for implementing the tool described in Appendix B, Dr. Xufei Wang, and Ali Abbasi for providing help in our preparation and revision of Chapter 4. The authors would like to express their deep gratitude to Dr. Hiroshi Motoda for his valuable and detailed comments. Without Diane Cerra's patience and encouragement, this work would not be possible to appear this year. This work was funded, in part, by the Office of Naval Research [ONR N000141110527], and the Army Research Office [ARO 025071]. Dr. Geoffrey Barbier was sponsored, in part, by the Office of the Secretary Defense-Test and Evaluation [OSD-T&E], DefenseWide/PE0601120D8Z National Defense Education Program (NDEP)/BA-1, Basic Research, and SMART Program Office[1] [N00244-09-1-0081].

Geoffrey Barbier, Zhuo Feng, Pritam Gundecha, and Huan Liu
April 2013

---

[1]www.asee.org/fellowships/smart

CHAPTER 1

Information Provenance in Social Media

1.1 SOCIAL MEDIA

Social media is defined in [39] as "a group of Internet-based applications that build on the ideological and technological foundations of Web 2.0, and that allow the creation and exchange of user-generated content." It is a conglomerate of different types of social media sites, including social networking (e.g., Facebook, LinkedIn, etc.), blogging (e.g., Huffington Post, Business Insider, Engadget, etc.), micro-blogging (e.g., Twitter, Tumblr, Plurk, etc.), wikis (e.g., Wikipedia, Wikitravel, Wikihow, etc.), social news (e.g., Digg, Slashdot, Reddit, etc.), social bookmarking (e.g., Delicious, StumbleUpon, etc.), media sharing (e.g., Youtube, Flickr, UstreamTV, etc.), opinion, reviews and ratings (e.g., Epinions, Yelp, Cnet, etc.), and community Q&A (e.g., Yahoo Answers, WikiAnswers, etc.). Table 1.1 shows some key characteristics of different types of social media [30].

Social media gives users an easy-to-use way to communicate and network with each other on an unprecedented scale and at rates unseen in traditional media, including newspaper, radio, and television. The most popular social networking site, Facebook, boasts over one billion users.[1] The number of Facebook users is more than three times the population of the United States, with 80% of Facebook users from outside of the United States,[2] and 500 million users from Nov. 2010 to Nov. 2012. Figure 1.1 illustrates the explosive growth of Facebook during its first eight years.[3] The popular microblog site Twitter has also experienced significant growth since its opening day in 2006 [51]. Twitter users send more than 400 million microblog messages, commonly referred to as "tweets," per day,[4] in 2012. In additional to sites like Facebook and Twitter, there are over 180 million web logs or "blogs" [37]. Between November 2010 and November 2011, the number of Facebook users increased by over a quarter billion, the number of mobile Facebook users increased by over 100 million, the number of tweets sent per day increased by over 90 million, and an additional 20 million blogs were added to the blogosphere. Additionally, social media sites are commonly used in the workplace, including government agencies. In the United States, federal government offices including the Internal Revenue Service,[5] the Department of Education,[6] and organizations like the

---

[1]http://newsroom.fb.com/News/457/One-Billion-People-on-Facebook, accessed on Dec 3, 2012.
[2]http://newsroom.fb.com/content/default.aspx?NewsAreaId=22, accessed on Feb 20, 2012.
[3]http://news.yahoo.com/number-active-users-facebook-over-230449748.html, accessed on Dec 2, 2012.
[4]http://www.mediabistro.com/alltwitter/twitter-400-million-tweets_b23744, accessed on Aug 2012.
[5]http://www.facebook.com/IRSRecruitment, accessed on Feb 20, 2012.
[6]http://www.facebook.com/ED.gov, accessed on Feb 20, 2012.

Table 1.1: Characteristics of different types of social media

Type	Characteristics
Online Social Networking	Online social networks are Web-based services that allow individuals and communities to connect with real-world friends and acquaintances online. Users interact with each other through status updates, comments, media sharing, messages, etc. (e.g., Facebook, LinkedIn, Myspace).
Blogging	A blog is a journal-like website for online users, a.k.a. bloggers, to contribute textual and multimedia content, arranged in reverse chronological order. Blogs are generally maintained by an individual or a community (e.g., Huffington Post, Business Insider, Engadget).
Micro-Blogging	Microblogs can be considered the same as blogs, but with limited content (e.g., Twitter, Tumblr, Plurk).
Wikis	A wiki is a collaborative editing environment that allows multiple users to develop web pages (e.g., Wikipedia, Wikitravel, Wikihow).
Social News	Social news sites allow a community of users to select and share news stories and articles (e.g., Digg, Slashdot, Reddit).
Social Bookmarking	Social bookmarking sites allow users to bookmark web content for storage, organization, and sharing (e.g, Delicious, StumbleUpon).
Media Sharing	Media sharing is an umbrella term that refers to the sharing of a variety of media on the web including video, audio, and photos (e.g., YouTube, Flickr, UstreamTV).
Opinion, Reviews, and Ratings	The primary function of such sites is to collect and publish user-submitted content in the form of subjective commentary on existing products, services, entertainment, business, places, etc. Some of these sites also provide product reviews (e.g., Epinions, Yelp, Cnet).
Community Q&A	These sites provide a platform for users seeking advice, guidance, or knowledge to ask questions. Other users from the community can answer these questions based on previous experiences, personal opinions, or from relevant research. Answers are generally judged using ratings and comments (e.g., Yahoo Answers, WikiAnswers).

Federal Emergency Management Agency (FEMA) are leveraging social media sites like Twitter as well.[7]

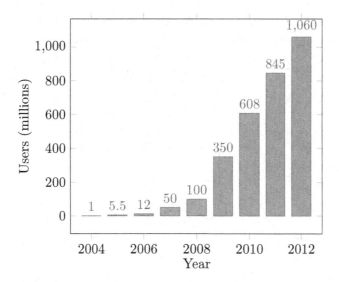

Figure 1.1: Numbers of Facebook users over the years.

In addition to providing a popular means to connect with friends, associates, and family members, social media has practical applications that are benefiting society as a whole. Social media has been used for gathering information about large-scale events such as fires, earthquakes, and other disasters, all of which impact government and non-government organizations at local, national, or even international levels [20, 21]. Individuals also use social media to find reliable information about what is going on around them and thus are able to leverage new information as quickly as possible.

One characteristic of social media is its low entry barrier enabling its wide use and explosive growth. Users simply need access to the Internet to participate in social media today. With the ubiquitous availability of computational resources and Internet access, people produce a variety of content and interact with many others directly through social media. This is vastly different from the traditional media such as radio, television, and printed publications that dominated social communications in the past. These traditional media mechanisms are available to individuals or organizations with sufficient financial means to purchase "air time" or space on a printed page. Traditional media convey messages in a one-to-many fashion.

Figure 1.2 illustrates how information is propagated as in some traditional media, in this example, a newspaper. Unlike social media, news and information is collected and prioritized by a relatively small subset of people (reporters), then prioritized and published in the form of a newspaper, and then distributed to the much larger set of people (readers). Readers providing feedback on

---

[7]http://twitter.com/fema, accessed on Feb 20, 2012.

information or opinions published in the newspaper are subject to the judgement of the newspaper editors and the time delay of publishing the response in a future edition of the newspaper.

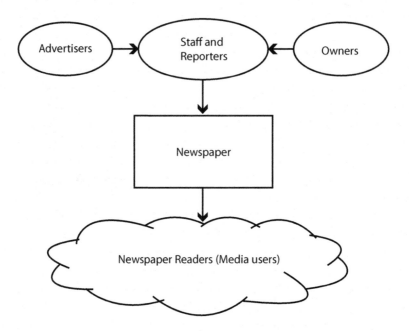

Figure 1.2: A newspaper (traditional media) framework for information sharing among its users.

Figure 1.3 illustrates how social media users share information. Unlike traditional media, an individual user in social media can transmit or retransmit information to and from other users without requiring feedback or checking the trustworthiness of information. Hence, social media allows the average user to reach more users instantly than they could before. Social media allows many-to-many communication among its users.

Social media has profoundly impacted the way people interact and communicate. Personal messages, opinions, news, and marketing material are all common uses of social media today. Social media propagates breaking news and rumors alike on an unsurpassed scale. Its broad use and rapid growth is impressive and raises many challenges. Tang and Liu [62] identified some tasks to address key social media challenges, including network modeling, centrality analysis and influence modeling, community detection, classification and recommendations, privacy, spam, and security.

1.2   SOCIAL MEDIA DATA

Social media data is largely user-generated, vast, noisy, distributed, unstructured, and dynamic in nature [30]. It is primarily available in the form of individual users' attributes, user-user connections (links), or user-generated content, including texts, photos, and videos (refer to Figure 1.3).

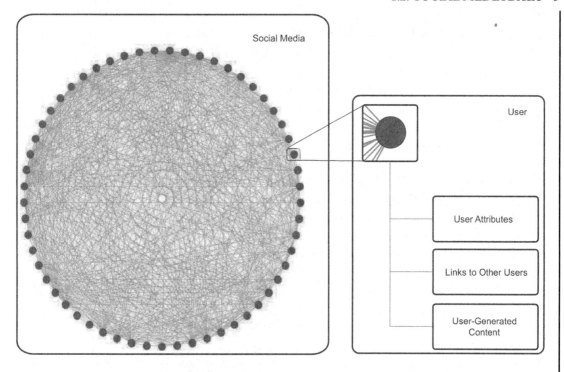

Figure 1.3: A social media illustration for information sharing among its users.

An individual user can share a large amount of personal and sometimes sensitive information with friends on a social networking site through the user's profile, status updates, messages, and status replies. Figure 1.4 shows a typical user profile on Facebook. Depending on the individual's choice, the user profile can reveal personal information such as gender, birth date, relationship status, e-mail address, phone number, home address, and even political or religious affiliations. The number of attributes contained in a profile varies widely, based on the social media site and user preference. Table 1.2 shows percentage of publicly visible attributes through Facebook users' profiles [28].

Besides user attributes, social media sites allow users to link to each other. These links are either undirected (e.g., Facebook) or directed (e.g., Twitter). Links can be indicative of friend or follower/following relationships, professional associations, group membership, family ties, etc. Around 70% of Facebook users publicly share all of their friend relationships [28]. Figure 1.5 is a graphical presentation of the first author's Facebook network. Each blue dot included in the figure is considered a node in the graph and represents one individual in the first author's social network (i.e., the author has a Facebook "friend" association with every individual shown in the graph). Each undirected link presented in the figure illustrates how two nodes associate with each other in the author's social network in addition to the "friend" association with the author.

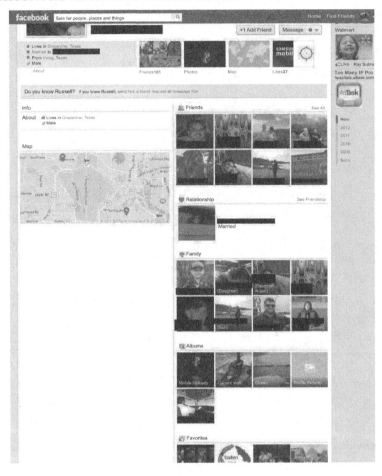

Figure 1.4: A typical user profile on Facebook. Attributes including name, gender, family information, hometown, current location, friends, and community affiliations are publicly shared with others.

Sometimes links in social media can be directed. For example, a user in Twitter generally connects to two sets of users: the follower set and the following set. The follower set of a user, represented by incoming edges, consists of those users who are subscribed to a user's tweets, whereas the following set of a user, represented by outgoing edges, consists of those users to whom a user is subscribed. Figure 1.6 is a graphical representation of a few Twitter users who were involved in a disinformation spread in "Assam Exodus" (see detail in Section 1.3) and their follower-following relationships. Each blue dot in the figure is considered a node in the graph and represents one individual in Twitter. Each directed link presented in the figure illustrates how two Twitter users associate with each other.

Table 1.2: Percentage of user attributes publicly visible through Facebook users' profiles

Attribute	Percentage (%)	Attribute	Percentage (%)
Current City	30.17	Hometown	35.38
Gender	81.77	Birthday	3.30
Relationship Status	26.24	Siblings	11.90
Education and Work	25.13	Like and Interests	66.57
Email	1.32	Mobile Number	0.36
Website	6.26	Home Address	0.37
Political Views	1.19	Religious Views	1.61
Children	4.21	Networks	13.83
Parents	3.49	Bio	9.68
Interested in	18.66	Looking For	21.86
Music	45.77	Books	13.68
Movies	27.92	Television	33.30
Activities	18.74	Interests	14.99

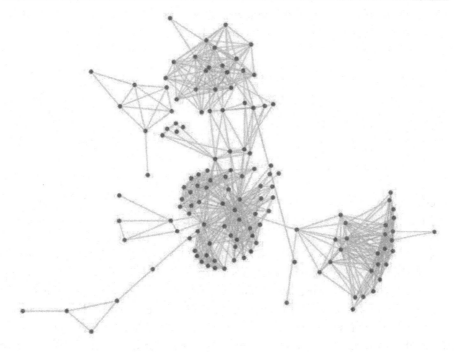

Figure 1.5: Undirected links representing friend relationships among members of the first author's Facebook network.

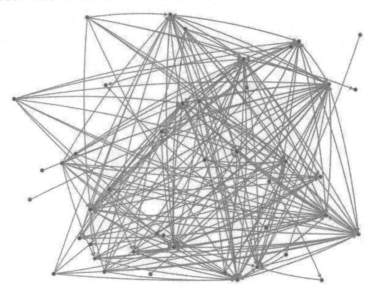

Figure 1.6: Directed links representing follower-following relationships among a few Twitter users who were involved in a disinformation spread in "Assam Exodus."

Attribute data and link data are not the only data found in social media. Social media users post massive amounts of user-generated content, including messages, comments, photographs, videos, and articles. For example, popular television networks ABC, NBC, and CBS, over the course of 60 years, produced 1.5 million hours of programming. Contrast that amount with YouTube,[8] a popular social media site, which received more video in six months than all three television networks produced in total during these 60 years [65].

Today, social media data can be propagated very quickly to a large number of users. One article even compares the speed of "Tweet Waves" to seismic waves [4]. This ability to propagate information quickly can be useful or, in some cases, harmful. For example, social media can be leveraged to assist with humanitarian aid and disaster relief [20], but social media can also serve as a mechanism for rumor propagation [2, 57]. If additional data related to a particular statement is made available to social media users, users can better discern whether a statement published in social media is reliable, or if it contains false information. Unfortunately, the existing structure of social media provides no mechanism allowing its users to inspect received information. Social media needs better mechanisms to understand what is being disseminated online. In most cases, users have no basic metadata about received information, e.g., the *provenance*, a.k.a. *sources* or *originators* of the information. The next section discusses what information provenance is and why it is important.

---

[8]http://www.youtube.com/

1.3   INFORMATION PROVENANCE

Traditionally, the provenance of an object informs its ownership, source, or origin. Provenance of a painting informs who the original artist is, thus can increase its value. Provenance is determined using its related information. For example, provenance of a painting is determined using an available history of ownership. Provenance of an experiment is determined using what activities preceded the experiment, what parameters were used in the experiment, and what the settings for the experiment were. This information can help scientists to understand successes and failures and even reproduce experimental results. In social media, provenance of information provides a similar value to its users. Social media contains social media data (attributes, links, and contents) which can be used to determine information provenance.

Knowing the provenance of a piece of information published in social media—how the piece of information was modified as it was propagated through social media and how an owner of the piece of information is connected to the transmission of the statement—provides additional context to the piece of information. A social media user can use this context to help assess how much value, trust, and validity should be placed on the information.

One of the important applications of information provenance in social media is to identify rumormongers or disinformation centers. The "Assam Exodus" [27] is a recent example that illustrates the importance of information provenance. Assam is a large state in north-east India where a series of riots broke out in July and August 2012. Following the riots, virulent messages, along with disinformation, were spread to other parts of India via social media. Bulk text messages (short message services, SMS) and social media sites were extensively used to spread information whose objective was to incite certain Indian populations against the north-east Indian population. For example, a Wall Street journalist reported that a Twitter user used a gory video clip of riots in Indonesia as that of the Assam riots.[9] Violent messages were also spread on Facebook that incited hatred and vengeance against the north-east Indian population. The disinformation, as well as the virulent messages, resulted in deep fear among the north-east Indian population, which ultimately led to their exodus from some major metropolitan cities across India, including Bangalore, Mumbai, Hyderabad, Chennai, and Pune.[10] In such cases, information provenance might be able to help find the rumormongers or disinformation centers early and help stop the viral spread of disinformation.

Information provenance also provides important context for assessing statements that are presented as fact or opinion. For example, in 2010, Twitter user *villaraigosa* published the following statement via Twitter: "MTA to pursue fed $ 4 Subway & Regional Connector! Projects that will cut pollution, create jobs and relieve traffic http://bit.ly/2vyBWK." Following the URL included in the message and examining *villaraigosa*'s user profile reveals that the statement was put forward by the mayor of Los Angeles, California. Knowing *villaraigosa* is the mayor adds context to the statement that a user can include in making judgments about the information conveyed.

---

[9]`https://twitter.com/dhume01/status/236321660184178688`, accessed on Oct, 2012.
[10]`http://en.wikipedia.org/wiki/2012_Assam_violence`, accessed on Oct, 2012.

The utility of information can be illustrated further with two earlier cases. In early 2010, it was rumored that the Chief Justice of the United States Supreme Court was going to retire due to medical reasons. In fact, the Justice had no plans to retire. The statement originated from a Georgetown University Law School class, and was meant only to be a teaching point. However, with the availability of the Internet, before the Law professor revealed the falsehood, students in the class had transmitted the statement, which was subsequently published on a news blog [9, 54]. Had the provenance been made available, recipient users might not have considered the statement credible. In another case, a United States Department of Agriculture employee was erroneously fired after information about her appearing in social media was published out of context [57]. Had information provenance been available, sought out, or examined, it might have prevented an injustice to the employee and embarrassment for the Department of Agriculture.

So far, we show what information provenance is and why it is necessary to identify it. In the next section, we formally introduce the information provenance problem.

1.4   THE INFORMATION PROVENANCE PROBLEM

A network can be represented as a directed graph $G(V, E, p)$, where V is the node set and E is the edge set. Each node $v \in V$ in the graph represents an entity, which can publish, receive, and propagate pieces of information in social media. An entity refers to an individual user (Facebook or Twitter user, blogger, etc.) or a webpage (blog or news article, YouTube video link, etc.). A directed edge, $(u \rightarrow v) \in E$, between nodes $u, v \in V$ represents the direction of information propagation. Each directed edge $(u \rightarrow v)$ is assumed to be associated with an information propagation probability: $p(u \rightarrow v) \in [0, 1]$. $p(u \rightarrow v) = 1$ means information always propagates from node u to node v, whereas $p(u \rightarrow v) = 0$ means information never propagates from node u to node v.

For information propagating through social media, provenance informs a user about its sources, $S \subseteq V$. Sources refer to the nodes that first publish the information. Starting from sources, all the nodes that are part of information propagation are referred to as *recipients*, $R \subseteq V$. Recipients can also propagate information, or retransmit it with modifications. During the propagation process, the information is transmitted through different nodes in social media, referred to as *transmitters* $M \subseteq R$. Recipients that do not propagate information further are referred to as *terminals*. In social media, it is quite challenging to obtain all the recipients of given information. In this work, we are interested in seeking the provenance of information from a few known terminals T, perspective, that are assumed to have received the information. Appendix A lists some important terms and symbols that are used throughout this lecture notes.

The public Facebook network dataset [45] in Figure 1.7 shows some nodes participated in information propagation. Each node represents a Facebook user. Each edge represents a friend relationship observed in the Facebook. Grey dots do not receive information, whereas the (big) green solid circle (upper right corner), the red solid circles and the blue solid triangles are the recipients of information. In this example, the green solid circle is the source where the given information propagation originates or starts. The transmitters other than the source, red solid circles, are the nodes

Figure 1.7: A Facebook network shows some nodes participated in information propagation. Each node represents a Facebook user. Each edge represents a friend relationship observed in the Facebook. Grey dots do not receive information, whereas the (big) green solid circle (upper right corner), the red solid circles, and the blue solid triangles are the recipients of information. The green solid circle is the source of a given information propagation. Red solid circles are the transmitters, whereas blue triangles are the terminals.

which successfully transmit or retransmit information. The terminals, blue solid triangles, are the nodes which receive but do not transmit information further. Many times, information propagation in social media is more complex than shown in Figure 1.7. For example, during the events of "Assam Exodus" and Hurricane Sandy, information containing violent and/or rumor content was often propagated by a large number users across multiple social media platforms.[11]

Given the above defined notations, the information provenance problem can be formally stated as below.

Problem 1.1   The INFORMATION PROVENANCE Problem   Given a directed graph, $G(V, E, p)$, with known terminals, $T \subseteq V$, and a positive integer constant, $k \in Z^+$, identify the sources, $S \subseteq V$, such that $|S| \le k$, and $U(S, T)$ is maximized. Function $U(S, T)$ estimates utility of information propagation that starts from the sources, S, and stops at the terminals, T.

$$\hat{S} = \underset{S \in V, |S| \le k}{\operatorname{argmax}} U(S, T), \tag{1.1}$$

where $\hat{S}$ represents estimated sources. The INFORMATION PROVENANCE problem aims to estimate $\hat{S}$ of the original sources S^*. Conventionally, estimation of utility function $U(S, T)$ is dependent on the underlying information propagation model, such as susceptible-infected, susceptible-infected-recovered, independent-cascade, or linear-threshold models.

Information in social media can originate from multiple sources. Therefore, the INFORMATION PROVENANCE problem requires a positive integer constant, $k \in Z^+$, as an input to identify the multiple sources for known terminals T. In the case of single source identification, setting $k \ge 1$ helps find more than 1 source to increase the likelihood of sources.

Social media can facilitate solving the information provenance problem due to its unique features: user-generated content (e.g., tweets, blog posts, news articles, etc.), user profiles, user interactions (e.g., links between friends, hyperlinks on blogs, or news articles), and spatial or temporal information. These features can help reconstruct a network with information propagation, which is essential for information provenance. In the next section, we discuss challenges to tackling the information provenance problem.

1.5   CHALLENGES

Common approaches for managing electronic provenance do not address contemporary social media. The Word Wide Web Consortium[12] (W3C) initiated a working group to provide recommendations for possible standards [16]. However, the underlying assumption is that there will be widespread use of the semantic web [24], or publicly available linked provenance data [33], which is not the case

---

[11]http://in.reuters.com/article/2012/08/21/sms-socialmedia-assam-migrants-idINDEE87K09120120821, accessed on Oct, 2012.
[12]http://www.w3.org/, accessed on May 25, 2012.

for social media. Thus, today's social media users will not be able to leverage such standards in the future unless the standards become widely used across social media sites such as Facebook, Twitter, LinkedIn, and Google+.

Outside of social media, the value of obtaining and tracking provenance is well understood in many computational applications, including databases [12, 14, 69], e-Science [58], and distributed processing [55]. Here, the provenance of information is required to determine the authenticity and trustworthiness of information, and solve data conflicts. The primary research focus in these areas is to redesign storage and management systems that can facilitate the provenance data later. These applications typically rely on some type of provenance storage, either centralized or distributed [23, 60]. For example, provenance data for a physics simulation may include parameters such as databases used and simulation settings used for the experiment. The advent of cloud computing has also brought approaches for implementing provenance storage [55]. However, social media sites have not implemented distributed or centralized stores for information provenance.

The W3C incubator group provided a list of provenance dimensions that are applicable to provenance data in social media. The report [16] documents a gap analysis for scenarios. One of the scenarios, "News Aggregator," lists the following challenges, also found in social media, which motivate an approach to mining social media for information provenance:

- *"No common format and application programmer's interface (API) to access and understand provenance information, whether explicitly indicated or implicitly determined."* Social media sites do not provide provenance data today.

- *"Developers rarely include provenance management or publish provenance records."*

- *"No widely accepted architecture solution to managing the scale of provenance records."* Searching for provenance data "on-demand" and in near real-time would help to reduce the need to maintain large provenance stores.

- *"No existing mechanisms for tying identity to objects or provenance traces."* The same challenge exists in social media, which is the motivation for developing methods (refer to Chapter 3) to seek provenance paths [7].

- *"Incompleteness of provenance records and the potential for errors and inconsistencies in a widely distributed and open setting such as the web."* This is also a challenge in the dynamic social media environment, where information is published rapidly, by many people simultaneously, and with different view points.

Addressing the information provenance problem tackles some of these challenges.

1.6 IN SEARCH OF PROVENANCE DATA

Our goal is to enable the assessment of the trustworthiness of a piece of information received by a user. Information provenance is one of the critical pieces required for this assessment. As discussed above, identifying information provenance itself is a challenging task. This section discusses the pertinent question of how to identify information provenance in social media. To this end, we outline some essential tasks.

1.6.1   ANALYZING PROVENANCE ATTRIBUTES

Provenance attributes of a user may include name, location, gender, occupation, political and religious affiliations, information content, and a list of potential recipients who might have played some role in transmitting information (e.g., retweets information in a tweet). These attributes could be vital to the task of identifying the provenance of information. Provenance attributes help to narrow down the possible sources and give more credibility to a piece of information. Barbier in his Ph.D. dissertation [8], shows that many attributes of a user can be collected from Twitter alone. The existing collection system [8] is extended by merging the provenance attributes result from multiple platforms, including Twitter, Facebook, Topsy,[13] LikedIn, Wikipedia, and top Google search results. In Chapter 2, we describe the mechanism that these attributes in provenance searching. Appendix B describes the collection tool for provenance attributes in detail.

1.6.2   SEEKING PROVENANCE VIA NETWORK INFORMATION

Network information is essential for seeking the provenance of information. There are two possible approaches. First, use the available information to directly seek the provenance. This approach is applied when all the recipients are known for a piece of information. The second approach is to find the information propagation flow from sources to known recipients, as close as to the actual sources; and, based on the propagation flow, identify sources. This approach can be used even if a few recipients are known. In Chapter 3, we discuss some methods for both approaches of seeking provenance.

1.6.3   SEARCHING FOR PROVENANCE DATA

The provenance attributes approach only uses content information, while the provenance paths approach only uses the network structure information to determine the sources. However, how to apply provenance attributes to guide a more accurate provenance seeking is an open challenge. Chapter 4 describes a framework to employ existing network information as well as attributes [8] and propagation history [63] to obtain the provenance of information. The framework is based on an iterative method, where every iteration makes use of the network, as well as attribute information or propagation history alternatively, to reduce the search space and guide the provenance search.

1.7   SUMMARY

We have witnessed social media growing at an unprecedented rate in the last few years. Characteristics of social media make possible various nefarious activities that could hinder its potential growth.

---

[13]Topsy provides a platform for searches of content published on Twitter and the Web.

Social media has been the target of rumors or disinformation spreading, privacy concerns, security breaches, and trust exploitation. In this chapter, we introduced the information provenance problem that presents a challenge and hope that, by addressing it, we can pave the way for solving many important issues such as source trustworthiness, information reliability, and user credibility.

Chapter 2 introduces the provenance attributes-based approach to access a piece of information. Network information alone provides opportunities to identify the provenance of information. Chapter 3 details the provenance paths problem, which only uses network information to determine information provenance. In Chapter 4, we describe the framework for searching provenance data and show later that the heuristic based on this framework aims to solve the provenance data problem using provenance paths as well as provenance attributes and propagation history together.

We hope that this lecture will help readers appreciate the problem of information provenance and the pressing need for effective solutions, understand challenges, present progress and huge potentials, build on preliminary accomplishments to develop novel methods toward solving the problem of information provenance, and facilitate the healthy development of social media.

CHAPTER 2

Provenance Attributes

When a social media user receives information via a microblog message, a social network, or even a blog site, it is not always clear where the received information originated from, what motivated its publication, and what latent purposes may be associated with it. In such circumstances, with additional metadata, a user could make a better-informed judgment about the received information. For example, when attributes such as displayed name, occupation, education level, or age can be associated with the originator of information, a user is better informed *about* received information. In a particular domain, such as politics, a user may be interested in additional pieces of metadata. For example, a user with political interests may add political affiliation and special interests to the list of desired attributes.

From the information provenance point of view, this chapter attempts to answer the following questions: What kind of metadata *about* the received information in social media is useful for a recipient to identify information provenance? How can we collect and measure the metadata qualitatively?

This chapter discusses provenance attributes, one of two key aspects of provenance information in social media. Definitions, metrics, and reviews of the preliminary analysis of provenance attributes in the context of one node of the social network will be presented. Provenance attributes are data of interest associated with a particular social media node; specifically, a social media node that published or propagated a statement of interest (information) on a social media network. In practice, sets of provenance attributes are defined subjectively, based on the particular interest of the information user (a recipient). As will be shown, some attribute values are easier to obtain than others and some attribute values may be more valuable to a recipient than others. For example, a political statement published by a political candidate might be assessed with some bias if the recipient knows information about the political candidate, such as political party affiliation or special interest associations. An interesting example of the value of provenance attribute data would be to reveal the political affiliation and special interests of an unfamiliar social media user propagating political statements, which may help understand latent motivations for propagating a statement in social media.

Table 2.1 displays a "general" and a "domain-specific" attribute list, the provenance attribute sets analyzed in [8]. The attribute sets in Table 2.1 are grounded in standard demographic information [11].

Specifying the particular set of provenance attributes of interest forms the foundation from which to begin the search for information provenance in social media (refer to Chapter 4). Formally defining provenance attributes provides a basis for us to define metrics for gauging progress in obtaining provenance attribute value data and, in turn, assists with analyzing provenance attributes.

Table 2.1: Lists of general and domain-specific provenance attributes.

General Demographic Attribute Set	Domain Specific (Political) Attribute Set
Formal Name (Individual or Group)	Formal Name (Individual or Group)
Location	Location
Occupation	Occupation
Education	Education
Age	Age
	Employer
	Political Affiliation
	Lobby Affiliation
	Special Interest(s)
	Conviction(s)
	Citizenship
	Ethnicity
	Gender

2.1   DEFINING PROVENANCE ATTRIBUTES

Measuring provenance attribute values that are not readily provided or trivially obtained provides new information to a recipient. Here are the formal definitions related to provenance attributes:

- α is a unique identifier, such as a username. There is an underlying assumption that each node in $G(V, E, p)$ is unique is some manner. α is an identifier that can be constructed to uniquely identify a node. A common example would be a user name that is unique to a particular social media service. For example, many domain user names for services such as Twitter and Google are unique.

- A is a set of provenance attributes, $(a_1, \ldots, a_n) \in A$, sought for any α. For example, provenance attributes might include *name, occupation, and education*. Provenance attributes are the metadata about a particular node that a recipient user is interested in. Note that provenance attributes are a preselected list of data elements that a user chooses, based on what is important and valued from the user's perspective.

- N is the number of provenance attributes sought after for any α. $N = |A|$. Enumerating the number of attribute values desired allows us to develop strategies for quantitative assessment criteria that can be used to distinguish between provenance attribute values, and by extension, nodes, users, and paths.

- W is the set of weights, $(w_1, \ldots, w_N) \in W$, associated with $(a_1, \ldots, a_N) \in A$. Weighting particular provenance attributes allows us to develop strategies for quantitative assessment criteria.

- $\mathfrak{V}_\alpha$ is the set of provenance attribute values, $(\mathfrak{v}_1, \ldots, \mathfrak{v}_N) \in \mathfrak{V}_\alpha$, associated with α. The collection of attribute values is meant to inform the recipient user about a particular node, such that the recipient user can better assess information originating from the node associated with the attribute values. Both the presence and the absence of specific attribute values can be informative to a recipient. For example, the attribute values might be *Mike, Graphic Artist, Bachelor of Arts-New York University*, and *unknown*.

With definitions for what we consider provenance data in social media to be, we can develop specific methods for quantifying how much provenance metadata is available for a given social media statement.

2.2   MEASURING PROVENANCE ATTRIBUTES

The availability function [8] objectively quantifies progress in obtaining attribute values. It is defined as:

Definition: *information provenance availability* function,
$r : \mathfrak{V}_\alpha \rightarrow [0, 1]$,

$r(\mathfrak{V}_\alpha) = \frac{\sum_{n=1}^{N} w_n \times x_n}{\sum_{n=1}^{N} w_n}$ where $x_n = 0$ if $\mathfrak{v}_n$ is unknown, otherwise $x_n = 1$.

The availability function describes how much provenance metadata is available for the statement of interest. The availability function allows a user to perform simple comparisons of search strategies that are employed to obtain provenance attributes. Additionally, the availability function allows a recipient to prioritize search results. For example, specific user applications designed to obtain provenance attributes can be compared based on the number of attribute values found.

Referencing multiple sources to determine whether or not an attribute value associated with α is consistent across the sources helps to validate the attribute value. For example, "villaraigosa" is associated with the name "Antonio Villaraigosa" on a Twitter profile and a Facebook profile. The occupation "mayor" is associated with the name, "Antonio Villaraigosa," in the Twitter profile, Facebook profile, and on a City of Los Angeles web page. In this case, "villaraigosa" is α and the provenance attribute, *name*, was validated by two separate sources associated with α. Political Affiliation was validated by four sources including two social medial sites and two web sites (noted as "URLs") as shown in Table 2.2.

Quantifying the number of sources that provide *the same attribute value* associated with α provides a validity value for the provenance attributes associated with a specific statement [8]. Dividing the total number of sources found (that provide the same attribute value for a particular attribute of interest) by the average total number of sources found for an attribute for other social media messages in the same particular domain indicates whether the provenance metadata validity

Table 2.2: An illustrative example of provenance attribute sources.

Attribute	Example Source(s)	Source Counter Value
Formal Name	Twitter, Facebook	2
Occupation	Twitter, Facebook, LinkedIn	3
Political Affiliation	Facebook, Google+, URLs	4
Education	Google+, Facebook	2

is above or below average. Specifically, we define a set of counters and an expected total count value as:

- $I_{\mathfrak{V}_\alpha}$ are attribute value *source counters*, $(i_1 \ldots i_N) \in I_{\mathfrak{V}_\alpha}$, for attribute values in the corresponding $\mathfrak{V}_\alpha$.

- c is the *expected total source count* for a particular set of provenance attributes in a particular domain, A.

A hypothetical set of attribute source counters for α is shown in Table 2.2. c is calculated by summing the average counter values for a particular domain of interest to a recipient. To illustrate how provenance attribute values might be assessed for accuracy as described later in this section, we will assume the average counter values for each attribute are 3, thus, $c = 12$. c is effectively a constant, but only a particular domain of interest. More importantly, c is a reference meant to present the user with a basis of comparison about the number of sources that are consistent with a particular attribute value for a specific search for provenance data. Ideally, the value of the legitimacy function would be greater than one, but in reality, may not always be greater than one. With c as a reference value for messages in a particular domain, legitimacy values of less than one can also be helpful, but must be interpreted subjectively. For example, legitimacy values of 0.80 may provide the user some level of confidence in the attribute values found. However, a legitimacy value 0.10 may reduce the user's confidence in the attribute values, but may yet be a valuable indication that less confidence should be placed on the accuracy of the social media statement under consideration.

Given the example that $c = 12$, we will say that the average of three sources are in agreement with each attribute value in $\mathfrak{V}_\alpha$. Thus, we can calculate $l(I_{\mathfrak{V}_\alpha})$.

$$l(I_{\mathfrak{V}_\alpha}) = \frac{\sum_{n=1}^{N} i_n}{c} = \frac{5+3+4+2}{12} = 1.17$$

We will call the attribute values legitimate when $l(I_{V_\alpha}) \geq 1$, for the attribute set, $\mathfrak{V}_\alpha$. More work needs to be done to determine values for c pertaining to any particular domains of interest, such as such as politics, news, and entertainment [8].

The following function is proposed to quantify whether or not the attribute values found are valid [8]:

- *Provenance attribute value legitimacy* function,
  $l : I_{\mathfrak{V}_\alpha} \to \mathbb{R}$,

$l(I_{\mathfrak{V}_\alpha}) = \frac{\sum_{n=1}^{N} i_n}{c}$, where i_n = source count for attribute n.

Problem Statement for Legitimacy Maximization: Given statement $\mathfrak{S}$, unique identifier α, provenance attribute values $\mathfrak{V}_\alpha$, and expected total source count c; find attribute values $\mathfrak{V}_\alpha$ to maximize information provenance legitimacy l.

When it is necessary to gather provenance attributes from disparate social media sites, one challenge that may arise is duplicate names. For example, suppose α corresponds to a user with the first name of "Tom" and the last name "Jones." When the search for attribute values extends to another social media site with several users with the name "Tom Jones," which "Tom Jones" should we associate with α. In some cases, additional attributes might be leveraged to identify the correct user. For example, address, education, profile pictures, or other provenance attributes and corresponding attribute values would match the most likely "Tom Jones." However, attribute values may be missing or not even the same between two social media sites. Another option for matching a duplicate name is to identify the most likely match by comparing the users' social networks.

The following definitions could be used to assess the probability of matching a duplicate name with a particular α, based on the social network or friends:

- F_α is the set of the names of α's *friends*.

- F_η is a *set of friend names* associated with one duplicate name identifier on another social media site.

- $p(F_\eta)$ is the *probability of the match* of F_η to F_α,

  $p : F_\eta \rightarrow [0, 1]$,

  $p(F_\eta) = \frac{|F_\eta|}{|F_\alpha|}$.

For example, suppose α has Twitter followers with names a, b, c, d, and e. $F_\alpha = a, b, c, d, e$. When the search extends from one site (say Twitter) to another social media site like Facebook, we look for the "Tom Jones" who has the most overlap with F_α. The *first* "Tom Jones" found on Facebook has friends b, d, e, thus, $F_\eta = b, d, e$, and,

$$p(F_\eta) = \frac{|\{b, d, e\}|}{|\{a, b, c, d, e\}|} = \frac{3}{5} = 0.60 .$$

As there are several profiles on Facebook with the name "Tom Jones," $p(F_\eta)$ is computed for each duplicate profile. The duplicate profile with the greatest overlap has the highest probability of being the relevant profile associated with α. The search for provenance attribute values can continue using the profile with the highest probability.

Ahsan and Shah present metrics for assessing electronic provenance [3]. Their metrics include: granularity, representation, format, scalability, data core-elements, completeness, accuracy, conformance, timeliness, accessibility, authority, and security. Some metrics are better defined than others and some metrics are more useful than others. Although Ahsan and Shah define timeliness, accessibility, authority, and security loosely, these concepts may prove valuable for assessing social media provenance attribute values.

The Ahsan and Shah implementation of the timeliness metric combines the age of a document, the frequency of use, and their provenance accuracy metric as follows [3]:

$$age = present\_year - publication\_year$$
$$frequency\_of\_use = \frac{times\_retrieved}{total\_records\_retrieved} \quad \text{(over a period of a year)}$$
$$Q_{currency} = Q_{accuracy} \times age \times frequency\_of\_use \; Q_{accuracy} ,$$

assigns a score of 1 for every 10% of the original data elements that can be regenerated based on the provenance data that is available. This particular definition of accuracy is valid for a computer simulation or series of simulation experiments, but does not correlate well to the problem space of provenance data in social media. For some applications, such as a computer simulation, provenance data would be intended to enable the duplication of experimental results. For our social media application, provenance data serves a different purpose such that it is not needed to recreate the social media statement, instead informs a user about a social media statement. However, timeliness of social media data can be defined more simply as [8]:

$$Q_{currency} = (current\_time - time\_provenance\_data\_created)/retrieved .$$

Redefining timeliness for social media data provides for more general use and better addresses environments where currency might be evaluated repeatedly over short time periods such as hours. This is more pertinent for today's social media environment. Another aspect of timeliness that is important to account for in provenance systems is the time required to gather provenance attributes of interest [5, 59, 69]. This is especially relevant to social media information. A provenance system that takes longer to gather provenance attribute values than the frequency at which the provenance attribute values are, or are likely to be updated, does not provide accurate or valuable provenance attribute values.

Accessibility is not clearly defined by Ahsan and Shah. However, it is a useful conceptual metric for provenance attribute values in social media. It is has been demonstrated that there are conditions such that the provenance attribute values a social media user seeks may not be available as readily as anticipated or desired. For example, in one investigation, provenance attributes that were identified as useful for a political domain, and would be routinely available through demographic survey data or census data, were not readily accessible in social media [8].

There are different conditions that might lead to poor accessibility. First, search mechanisms may not be able to access attribute values because of information assurance, intellectual property, or privacy controls that are put in place by social media sites. Second, the social media sites may not collect and/or store attribute values. Finally, users simply may not publish the attribute values as part of their profiles.

Accessibility is different than the *Provenance Availability* function, which indicates how much provenance metadata was found for a particular statement of interest. Accessibility would measure whether or not an approach is able to obtain particular provenance attributes for a domain of statements. Accessibility could be a simple percentage of provenance attributes that can be accessed. When paired with particular provenance attribute search mechanisms, it would provide a factor for comparison between mechanisms.

Authority is enumerated but was not defined by Ahsan and Shah in [3]. Defining an authority metric does have utility for provenance attributes in social media. In previous work [8], a simple provenance engine was implemented to gather provenance attribute values in social media as a proof of concept. In retrospect, the social media sites were searched beginning with the site that was assumed to be the most accurate source of provenance attribute values. For example, after searching Twitter for provenance attribute values, the social networking site LinkedIn was the next site visited to search for additional attribute values. LinkedIn was chosen based on the accepted assumption that provenance attribute values identified in a public LinkedIn profile were more likely to be accurate than provenance attributes identified in a public Facebook or MySpace profile because, LinkedIn users are typically motivated to use LinkedIn for professional networking and career advancement. In this example, provenance attribute values obtained from LinkedIn could be characterized as having more authority than provenance attribute values for the same provenance attribute for the same α obtained from another site (defined to have less authority). With this approach, authority can be subjectively defined but objectively measured and compared, and implemented as a mechanism to distinguish the quality of provenance attribute values akin to Ahsan and Shah's motivation for enumerating "authority" as a provenance attribute metric.

A provenance security metric is also enumerated and not well defined in [3]. It can be important to ensure that provenance attribute values themselves are protected [17, 34, 47]. A simple metric based on a list of security features that a provenance system provides could be used with the sum of the number of features implemented to provide a usable metric [8].

Another aspect related to security could inform reliability or authority. For example, provenance attribute values obtained from users with secure profiles based on indexes like those described by Gundecha et al. [28] could be indicative of provenance attribute values that are more "authoritative," based on the hypothesis that users with secure profiles are more likely to have accurate profiles. However, this hypothesis needs to be tested.

2.3 ANALYZING PROVENANCE ATTRIBUTES

In the previous sections of this chapter, we formally defined provenance attributes and presented measurement methods for assessing them. In this section, we briefly highlight analysis provenance attributes. First author's research [8] investigated provenance attributes that could be associated with general social media users and explored an expanded set of provenance attributes for a specific domain, politics. Two methods were used to obtain provenance attributes, a manual method and an automated method. This section discusses and analyzes the two approaches. The manual analysis provided valuable insights into the challenges and opportunities of using social media itself to provide provenance attributes about received information in social media. Based on lessons learned during manual analysis, an automated approach was developed and implemented.

The primary motivation for manual analysis is to identify effective approaches for gathering provenance attribute values that can be automated in order to improve (i.e., shorten) the amount of time needed to gather them. Provenance attribute value is most valuable when it can be presented simultaneously with the social media information of interest. Although manual analysis can provide useful data on a small scale, it is desirable to automate the search and collection for provenance attribute values to address the scale, complexities, and opportunities afforded by social media. The ability to mine social media data itself for provenance attribute values presents opportunities to access metadata that otherwise would not be available in practice.

The strategy behind the manual search is to begin collecting provenance attribute values from the profile on the social media site with the originating message and search for other attributes values through other online sources, beginning with the most reputable sources. For example, recall our Twitter user "villaraigosa." On November 5, 2012, "villaraigosa" tweeted the following message: "#LA—Don't forget to #Vote tomorrow! You can find your local polling place & sample ballot here → http://bit.ly/QiAfZU #Election2012"

What might the motivation be for "villaraigosa" to send this message? Let's perform a manual analysis (search) for provenance attribute values and see what insights the provenance attribute values may provide. In this example α is the Twitter user name, so we begin with the Twitter user profile associated with α. From the Twitter profile summary[1] we obtain two provenance attributes and discover a *latent provenance attribute*. *Latent provenance attributes* are attributes that are not explicitly specified by the recipient, but can be leveraged to determine explicit provenance attribute values [8], from the Twitter profile.[2] In this particular case, we obtain the formal name, occupation, and location associated with α. The latent attribute[3] is a URL for $\alpha's$ web page.

The URL is a link to $\alpha's$ public Facebook page.[4] Information found at the URL validates $\alpha's$ formal name, location, and occupation. The attribute values found on the web page supplements the formal name value with a middle initial. Four additional provenance attribute values and another latent provenance attribute (another URL) are found by selecting the "About" link on the Facebook page.

The URL on $\alpha's$ Facebook profile page is a link to a web page associated with α.[5] Information found at the URL validates $\alpha's$ formal name, location, occupation, education, and special interests.

---

[1] https://twitter.com/villaraigosa/, referenced January 9, 2013.

[2] Research has revealed that the some users will include a variety of information about themselves in their Twitter profile page. The information on the profile page served as the starting point for the manual search in [8]. To the surprise of one researcher, some Twitter profiles contained detailed information including age, names of relatives, employer information, and even ages of relatives. Thus, there is a wide variety of data to support provenance attributes of interest to a recipient [8].

[3] Latent provenance attributes may not be consistently available, but should be considered when available.

[4] https://www.facebook.com/AntonioVillaraigosa, referenced January 5, 2013.

[5] http://www.mayor.lacity.org/index.htm, referenced January 9, 2013.

Searching LinkedIn public profiles using $\alpha's$ formal name[6] leads to another profile page for α that is a second validation of formal name, location, occupation, and education, and expands the list of $\alpha's$ interest areas. Additionally, all of the profile pictures that are associated with α at the three social media sites explored thus far (Twitter, Facebook, and LinkedIn) match. We now have some degree of confidence in $\alpha's$ formal name and some provenance attributes. The search for missing provenance attribute elements can be expanded to other social media sites and the Web. For example, a link to Wikipedia[7] supplements provenance attribute values with ethnicity and, by extrapolation,[8] citizenship.

The manual analysis associated with α reveals that the tweet originated from an active political candidate, and includes the candidates political party affiliation, citizenship, special interests, and convictions. With the provenance attribute metadata that is available from the manual analysis, a recipient can better assess $\alpha's$ genuine motivations for sending a message. This can be important in cases where the social media information can have real-life impacts on individuals and societies.

The results of the manual analysis of the provenance attribute values that correspond to the attribute sets included in Table 2.1 are presented in Table 2.3.[9]

Some users will publish more data in profiles than others. Public figures such as Mayor Villaraigosa are likely to publish more provenance attribute values than mainstream social media users. However, keep in mind that recipients may have access to additional profile data when able to access social media sites via their own credentialed login accounts.

The information provenance availability function presented in the second section of this chapter can be used to analyze how many provenance attributes are associated with a particular statement. Previous research [8] highlights the need for assigning weights, W, to A for a set of provenance attributes, $(a_1 \ldots a_n) \in A$, any α.

Recall that the availability function is a summary metric that describes the amount of provenance metadata available for a particular statement published in social media. Let us examine the availability function on the context of our example α, "villaraigosa." The availability function allows a recipient to perform comparisons of search methods that are employed to find provenance attribute values and to prioritize search results [8].

Recall the example tweet sent by α: "#LA—Don't forget to #Vote tomorrow! You can find your local polling place & sample ballot here → http://bit.ly/QiAfZU #Election2012." Thus, values for I and α are:

I = "#LA—Don't forget to #Vote tomorrow! You can find your local polling place & sample ballot here → http://bit.ly/QiAfZU #Election2012"

α = "villaraigosa."

---

[6]Referenced January 9, 2013.

[7]http://http://en.wikipedia.org/wiki/Antonio_Villaraigosa, referenced on January 9, 2013.

[8]I.e., he was born in the United States.

[9]Some attribute values were validated by multiple sources during the manual search. However, only the source of the initial value is listed in Table 2.3.

Table 2.3: Example of provenance attribute values found with manual analysis

Attribute	Value	Source
Formal Name	Antonio R. Villaraigosa	Twitter profile
Location	Los Angeles, CA	Twitter profile
Occupation	Mayor	Twitter profile
Education	UCLA '77	Facebook profile
Age	60	Web page
Employer	City of Los Angeles	Facebook profile
Political Affiliation	Democratic Party	Facebook profile
Lobby Affiliation	Multiple references	Web news articles
Special Interest(s)	Service Employees International Union...	Web page
Conviction(s)	Roman Catholic	Facebook profile
Citizenship	United States	Wikipedia
Ethnicity	Hispanic	Wikipedia
Gender	Male	Facebook profile

Table 2.4: Example of latent provenance attributes and values found with manual analysis

Latent Attribute	Value	Source
URL	http://www.facebook.com/AntonioVillaraigosa	Twitter profile
URL	http://www.mayor.lacity.org	Facebook profile

The value generated by the availability functions changes as the manual analysis progresses from site to site. We can calculate the availability function as the search progresses for the attribute values specified in Table 2.3. In this case, $A = \{$Formal Name, Location, Occupation, Education, Age, Employer, Political Affiliation, Special Interest(s), Conviction(s), Citizenship, Ethnicity, and Gender$\}$. Thus, $N = 13$. We will weight the provenance attribute, "Political Affiliation," equal to Formal Name. Thus, the weights are: $W = (100, 50, 50, 50, 50, 50, 100, 50, 50, 50, 50, 50, 50)$.

We will assess I from a provenance attribute perspective. The provenance attributes desired are A. As shown previously, values for all of the provenance attributes of interest are not found in the tweet or in $\alpha's$ Twitter profile. However, some attribute values were found referencing the Twitter profile, and we can calculate a value for the provenance attribute availability function as follows:

The provenance attribute available from Twitter results are formally $\mathfrak{V}_\alpha = ($Antonio R. Villaraigosa; Los Angeles, CA; Mayor; unknown; unknown; unknown; unknown; unknown; unknown; unknown; unknown; unknown; unknown$)$. Thus,

$$r(\mathfrak{V}_\alpha) \;=\; ((100\mathrm{x}1) + (50\mathrm{x}1) + (50\mathrm{x}1) + (50\mathrm{x}0) + (50\mathrm{x}0) + (50\mathrm{x}0) + (100\mathrm{x}0)$$
$$+(50\mathrm{x}0) + (50\mathrm{x}0) + (50\mathrm{x}0) + (50\mathrm{x}0) + (50\mathrm{x}0) + (50\mathrm{x}0))$$
$$/(100 + 50 + 50 + 50 + 50 + 50 + 100 + 50 + 50 + 50 + 50 + 50 + 50)$$
$$=\; 200/750 = 0.27\ .$$

With only the information provenance attribute values obtained from the Twitter profile page, the provenance availability of the tweet is 0.27. Extending the search to $\alpha's$ Facebook profile supplements the attribute values, resulting in $\mathfrak{V}_\alpha$ = (Antonio R. Villaraigosa; Los Angeles, CA; Mayor; UCLA âŁ™77; unknown; City of Los Angeles; Democratic Party; unknown; unknown; Roman Catholic; unknown; unknown; Male). Consequently, we can compute a new value for the provenance availability function:

$$r(\mathfrak{V}_\alpha) \;=\; ((100\mathrm{x}1) + (50\mathrm{x}1) + (50\mathrm{x}1) + (50\mathrm{x}1) + (50\mathrm{x}0) + (50\mathrm{x}1) + (100\mathrm{x}1)$$
$$+(50\mathrm{x}0) + (50\mathrm{x}0) + (50\mathrm{x}1) + (50\mathrm{x}0) + (50\mathrm{x}0) + (50\mathrm{x}1))$$
$$/(100 + 50 + 50 + 50 + 50 + 50 + 100 + 50 + 50 + 50 + 50 + 50 + 50)$$
$$=\; 500/750 = 0.67\ .$$

Recall that, with the provenance attribute values identified from the search of social media sites, values for all of the provenance attributes of interest were identified, reference Table 2.3. With a complete set of attribute values, the provenance availability value is computed to equal 1, as shown in the following update to the example:

> $\mathfrak{V}_\alpha$ = (Antonio R. Villaraigosa; Los Angeles, CA; Mayor; UCLA 77; 60;
> City of Los Angeles; Democratic Party; *Multiple references*;
> Service Employees International Union...;
> Roman Catholic; United States; Wikipedia; Male).

Consequently, we can compute a new value for the provenance availability function:

$$r(\mathfrak{V}_\alpha) \;=\; ((100\mathrm{x}1) + (50\mathrm{x}1) + (50\mathrm{x}1) + (50\mathrm{x}1) + (50\mathrm{x}1) + (50\mathrm{x}1) + (100\mathrm{x}1)$$
$$+(50\mathrm{x}1) + (50\mathrm{x}1) + (50\mathrm{x}1) + (50\mathrm{x}1) + (50\mathrm{x}1) + (50\mathrm{x}1))$$
$$/(100 + 50 + 50 + 50 + 50 + 50 + 100 + 50 + 50 + 50 + 50 + 50 + 50)$$
$$=\; 750/750 = 1.00\ .$$

The provenance attribute availability function provides a qualitative value to summarize how many, and how many important, provenance attribute values are available for I. The function accounts for variations in how important attributes are from each other by weighting each attribute. The greater the number of provenance attribute values available to a recipient, the better a recipient can rely on the provenance search to help assess I [8].

Although not needed in our example, manually extending a search from one social media site to another, such as from Twitter to Facebook to LinkedIn, may require duplicate names to be resolved. Previous research [8] leveraged location and profile photos (if available) to manually match a user on one site that corresponded with α to a profile on another social media site.

One author's manual search for provenance attributes provided some interesting insights into the problem space (some listed from [8]).

- There was more data in Twitter profiles than anticipated for some users. For example, some users listed age and political preferences. At least one attribute value, formal name, was identified for all of the tweets investigated in the manual analysis for the general attribute set.

- Not as many political attribute values were obtained manually as anticipated. Only about 37 of 150 of all the desired political affiliation attribute values were obtained.

- The URL listed for some users associated with their profile was useful in some instances (more so than the URL in the message).

- Social media profiles were easier to search as a logged-in site user (i.e., publicly available profile pages did not provide as much of the desired data as thought possible.) Matches for some individuals were realized by manually matching the profile pictures between different social media sites (i.e., resolving the entity resolution problem for some individuals). It is likely that automatically matching profile pictures would prove more challenging, but, in some cases, a profile picture is the same across social media sites.

- Manual web search proved very useful by providing links to sites with additional profile data, including social networking sites, blog posts, and personal web sites.

- Politicians appear to be more public about political attributes.

- Handling duplicate identities is a challenge that must be overcome. Manually, images combined with provenance attributes adequately addressed this challenge. Automated means for dealing with duplicate identities are needed in order to enable effective automated search capabilities. Comparing friend networks of social media users to identify similarities may help (i.e., social networks containing some identical user nodes might lead to identifying the same user relationships across social media sites). Advanced approaches to identity resolution, like the techniques developed by Jeff Jonas [38], might also prove useful in this area.

- The ability to identify some common provenance attribute values is likely not dependent upon a particular domain. For example, provenance attributes such as Formal Name, Education, and Location are often expressed in user profiles. Some domain-specific attribute values may be difficult to obtain because of privacy practices, security policies, and user choice. There may even be value in attribute values that cannot be found. Consider that in earlier research, attribute values for ethnicity, citizenship, and lobby affiliation were extremely rare finds. When a rare attribute value is found, the provenance attribute provides a greater amount of information to the recipient.

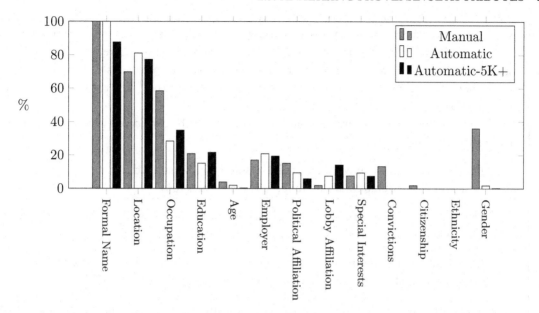

Figure 2.1: Comparison of percentages between manual and automatic search of political attributes to include over 5,000 α identifiers.

- It is noted that the Twitter biography may be a valuable provenance attribute in and of itself. It has been observed that the biography can provide a significant number of provenance attributes including age, occupation, employer, political affiliation, and interests. The biography may include insights about opinion, attitude, and sentiment that are best interpreted by a human recipient-user. Including the entire biography as a provenance attribute might be valuable to recipients and, because of the text field limitations placed on the biography by Twitter, not overly burdensome for storage and processing. In the end, the biography may serve both as a source of provenance attribute values and as an attribute.

Figure 2.1 presents the results of an automated search for provenance attributes for over 5,000 Twitter user names[10] (i.e., bars corresponding with "Automatic-5K+") [8]. Figure 2.1 also includes results of manual and automated (i.e., bars corresponding to "Manual" and "Automated") search for provenance attributes for user names associated with 150 interesting tweets. The same set of tweets is used in "Manual" and "Automated" searches. By contrast, "Automatic-5K+" results illustrate how simple automated approaches scaled to a relatively large number of users, and highlights challenges to obtaining some attribute values automatically, such as gender and occupation.

Note that, in the case of a politically motivated collection of provenance information in social media, it can be very challenging to find some valuable attributes, including: political affiliation,

[10]The Twitter user name, α, is used as the unique identifier.

lobby affiliation, special interests, convictions, citizenship, ethnicity, and gender. For some valuable provenance, no data values might be returned using automated means. Although it can be argued that more sophisticated search applications could be employed in [8], it is apparent that some attribute values are more easily accessible than others in social media. Thus, attributes that are insightful and scarce should be weighted as more valuable than others when basic[11] provenance attribute values for a given α are coincident.

Looking forward, there are key challenges that need to be considered when analyzing provenance attributes, including:

- Reconciling α across provenance attribute sources in social media due to the different user names a user may employ at different social media sites [66].

- Name-entity resolution; one challenge is to deal with situations when one social media user may have the same formal name as another social media user [35].

- Use of false names by social media users and the challenge this presents to discovery of provenance attributes associated with the correct individual.

- Sparsity of provenance attribute values collected from publicly available profiles compared to provenance attribute that would be available from a user's own social network.

These challenges provide rich opportunities for future research.

2.4   SUMMARY

Provenance attributes and associated provenance attribute values provide information *about* information, I, appearing in social media. Attributes, motivated by the subjective interests of a recipient, can provide deeper insights and context about information in social media. Analysis of social media sites can provide beneficial provenance attribute values that can better inform recipients about latent motivations and meanings associated with published information in social media. An automated tool for obtaining provenance attribute values [31] is developed and described in Appendix B.

---

[11] At minimum, the basic set should include the identity (i.e., α and name) of a social media user or source.

CHAPTER 3

Provenance via Network Information

In social media, information is often transmitted and retransmitted from one user to other users, and from one social media site to other social media sites. Although the existing structure of social media allows users to easily create, receive, and propagate a piece of information, it provides no mechanism for its users to know more about the received information, e. g., *provenance* (also known as, *sources* or *originators*) of such information. Previous research in social media shows that network information can be useful for many social media mining tasks, including community detection, network modeling, influence modeling, classification and recommendation, and privacy, trust, and security [18, 50, 62]. Network information is instrumental in seeking the provenance of information. In this chapter, we focus on the problem of *seeking the provenance of information in social media using network information*. We discuss two approaches to solving this problem. The first one is to use the available information to directly seek the provenance of information. This approach assumes that all the recipients are known for a piece of information. The second one is to find the reverse flows of information propagation, i.e., from the known terminals to sources. We refer to the second approach as *seeking provenance paths* [32].

The problem of seeking information provenance has received little attention in comparison with its counterpart, the study of information propagation. Information propagation refers to the spreading of information from one node to others in a network. Previous research has mostly focused on designing information propagation models [6, 13, 25, 26, 40, 49] with specific goals. For example, threshold and cascade models [40] of information propagation aim to identify the influentials in a network, whereas susceptible-infected (SI) based models [6] of information propagation aim to understand epidemics in a networked population. Information propagation models capture the characteristics of information propagation from sources to terminals. However, those models are insufficient to model the information provenance in social media that seeks the path from terminals to sources.

The rest of this chapter is arranged as follows. In Section 3.1, we present the information propagation models commonly assumed to solve the information provenance problem. In Section 3.2, we discuss two methods of directly seeking the provenance of information assuming that all the recipients are known, and one method of seeking provenance paths which we later used to determine the provenance of information if we only know a few recipients. We highlight different issues in seeking the provenance of information in Section 3.3.

3.1 INFORMATION PROPAGATION MODELS

The problem of seeking provenance of information assumes underlying information propagation models. Shah and Zaman [56] proposed a centrality based measure, called *rumor-centrality*, to identify the single-source node of a given rumor spread, based on Susceptible Infected (SI) model. Lappas et al. [43] proposed a method to estimate the multiple *effector* nodes of a given information spread under the assumption of Independent Cascade (IC) model. Prakash et al. [53] also proposed a method, called NETSLEUTH, to estimate the multiple sources under the assumption of the SI model. The susceptible-infected (SI) and independent-cascade (IC) models are the models most commonly used to address the problem of INFORMATION PROVENANCE. In the following section, we introduce the SI and IC models.

3.1.1 SUSCEPTIBLE-INFECTED (SI) MODEL

The SI model [6] is one most basic epidemic model. In this model, every node in the graph is in one of two states: susceptible (S) or infected (I). $S(t)$ represents the number of individuals not yet infected but susceptible to the disease at time t. $I(t)$ denotes the number of individuals who have been infected with the disease and are capable of spreading it to those in the susceptible state. Under the assumption of a fixed population, $N = S(t) + I(t)$. In the SI model, each infected node tries to infect each of its neighbors independently with probability β in each discrete time-step, which reflects the strength of the disease spread. Once a node is infected, it remains infected forever. In our context of information propagation in social media, recipients are in the infected state, whereas all other nodes are in the susceptible state.

Consider the network shown in Figure 3.1 applying the SI model. We assume that each infected node tries to infect each of its neighbors independently with probability β in each discrete time-step, i.e., $p(u \rightarrow v) = \beta$. If node 1 is infected initially, Figure 3.1 shows the information propagation process following the SI model. Starting from the initial stage with node 1 being active, the SI model chooses neighbors and activates them with propagation probability β. At time-step 1, node 1 tries to activate nodes 2 and 3. Suppose activation succeeds for node 3, but fails for node 2. Next, infected nodes 1 and 3 will try to activate neighboring susceptible nodes 2, 4, and 6. Say, nodes 2 and 6 become infected, while node 4 remains susceptible to infection at the end of time-step 2. At time-step 3, the infected set containing nodes 1, 2, 3, and 6 tries to activate susceptible neighboring nodes, including nodes 4, 5, and 7. Assume that node 4 becomes infected at this time, while the others remain susceptible. In the next time-step, the infected node set containing nodes 1, 2, 3, 4 and 6 aim to activate the remaining susceptible nodes 5 and 7. None of them get infected at time-step 4. At time-step 5, the same process, as in previous time-step 4, repeats, but results in the infection of nodes 5 and 7. Since no more susceptible nodes are left to infect, the propagation process stops. Note that, though the SI model infects each node with a certain success rates, the propagation process results in the infection of all the nodes (as long as they are reachable from the sources, in this case node 1).

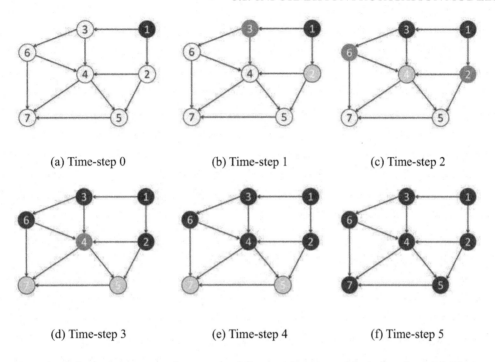

(a) Time-step 0   (b) Time-step 1   (c) Time-step 2

(d) Time-step 3   (e) Time-step 4   (f) Time-step 5

Figure 3.1: An information propagation process following the susceptible-infected model. Black nodes are the infected nodes, dark grey nodes are the newly infected nodes, light grey nodes with black outlines are the potential infected nodes in this step, and white nodes are those that do not get infected at this time.

3.1.2   INDEPENDENT-CASCADE (IC) MODEL

The Independent Cascade (IC) model [25, 26, 40] is a conceptually simple and widely adopted cascade model. The IC model is a probabilistic propagation model, where each node is assumed to be in one of two states: active or inactive. For a given directed graph $G = (V, E, p)$, the activation process starts with the source set (initial active nodes) $S \subset V$, and, following a randomized process, unfolds in a discrete number of steps. When node u becomes active at step t, it receives a single chance to activate each currently inactive neighbor v through the edge $(u \rightarrow v)$. Node u succeeds in this activation with probability $p(u \rightarrow v)$. If u succeeds, then v will become active at step $t + 1$. Otherwise, u is not allowed to make any more attempts to activate v in subsequent rounds. Note that if v has multiple, newly activated neighbors, then those active neighbors can independently attempt to activate v in any arbitrary order. This process runs until no more activations are possible. In our context of information propagation in social media, recipients are in the active state, whereas all other nodes are in the inactive state.

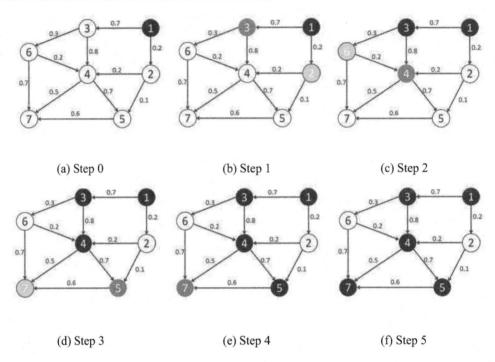

(a) Step 0   (b) Step 1   (c) Step 2

(d) Step 3   (e) Step 4   (f) Step 5

Figure 3.2: An information propagation process following the independent-cascade model. Black nodes are the infected nodes, dark grey nodes are the newly infected nodes, light grey nodes with black outlines are those whose activation are unsuccessful, and white nodes are those that do not get infected at this time.

Consider the network shown in Figure 3.2 applying the IC model. Each edge is labeled with its probability of successful information propagation. If node 1 is activated initially, Figure 3.2 shows the information propagation process following the IC model. Starting from the initial stage with node 1 being active, the IC model chooses neighbors and activates them with the propagation probability marked on the edges. At Step 1, node 1 tries to activate nodes 2 and 3. Suppose activation succeeds for node 3, but fails for node 2. Now, in Step 2, given newly activated node 3, we activate nodes 4 and 6 with probability 0.8 and 0.3, respectively. Say, node 4 becomes active and node 6 fails to become active. At Step 3, the newly activated node 4 tries to activate its inactive neighbors, including nodes 5 and 7. Assume that node 5 becomes activated successfully while all others fail to become active. Next, we consider node 5 trying to activate only inactive neighbor 7. Assume that this results in activating node 7. At Step 5, since no more activable nodes are left, the propagation process stops with nodes 1, 3, 4, 5, and 7 being active. Note that the IC model is a probabilistic model which activates a node with a certain success rate. Thus, we might get different output with the same initial state.

Table 3.1: Input assumptions by different methods of seeking the information provenance

Methods	Propagation Model	Multiple Sources	Directed Graph	No-Prior Knowledge of all Recipients
Rumor-centrality [54]	SI	No	No	No
Effectors [41]	IC	Yes	No	No
Net Sleuth [51]	SI	Yes	No	No
Pro Paths [29]	IC	Yes	Yes	Yes

Clearly, both the SI and IC models capture information propagation in a certain aspect and demonstrate significant differences. The SI model eventually ends up infecting all the nodes reachable from the sources of propagation. For the SI model, the amount of time required to spread infection depends on the number of infection sources at the start and infection probability β (the strength of infection). And the time of infecting all nodes is linearly related to the size of network. On the other hand, the IC model, even for two iterations with the same sources, may end up activating different sets of nodes.

Both the SI and IC models can also be applied to an undirected graph without any change, since infected or active nodes cannot be re-infected or re-activated. Other popular information propagation models include the linear threshold (LT) model [40], the susceptible-infected-recovered (SIR) model [49], and the susceptible-infected-susceptible (SIS) model [13]. These models follow relatively complex information propagation processes, in comparison with the SI and IC models.

3.2   SEEKING PROVENANCE OF INFORMATION

In this section, we present methods for seeking provenance of information. There are two approaches to solve this problem, based on the availability of recipient information. First, if we know all the recipients who have received the piece of information, then it is possible to use the available network information to directly seek the provenance of information. Otherwise, if we only know a few recipients, we find information propagation flow from sources to known terminals, as close as to the actual sources, then identify the provenance of information. We refer to the second approach as *seeking provenance paths*.

Based on the literature, the SI and IC models are the most preferred models to seek the provenance of information. Table 3.1 lists methods that aim to seek the provenance of information under different assumptions. Shah and Zaman [56] proposed a centrality based measure, called *rumor-centrality*, to identify the single-source node of a given rumor spread with all recipients known a priori. This work is based on a hypothesis that the source is at the center of the entire rumor spread and propagates information based on the Susceptible Infected (SI) model. Lappas et al. [43] proposed a method to estimate the multiple *effector* nodes of a given information spread with all recipients are known, under the assumption of Independent Cascade (IC) model. Effectors are recipients such that had the propagation started from them, it would have caused an information reception state similar to the one observed. Prakash et al. [53] also proposed a method, called NETSLEUTH, to estimate the multiple sources of a given information spread with all recipients known, under the assumption of SI model. However, this method automatically determines the number of sources required for a given spread. All of these methods are biased toward high-degree nodes, assume undirected network, and aim to directly seek the provenance of information. In contrast, Gundecha et. al [29] assume a directed network, as information propagation probability from user u to user v is not the same as that from user v to user u. Also, the proposed method requires knowing a few terminals (less than 1% of the total recipients) to start seeking the provenance paths.

3.2.1 DIRECTLY SEEKING SOURCES WHEN ALL RECIPIENTS ARE KNOWN

In this section, we describe two representative methods. One method focuses on finding a single-source and the other one is able to find multiple sources.

Identifying the Single Source

Shah and Zaman [56] propose a centrality based measure, called *rumor-centrality*, to estimate the single-source node, $k = 1$, of a given rumor spread with all recipients ($T = R$) known. This work is based on a hypothesis that the most likely source is at the center of the network and propagates the information based on the SI model. The graph G is assumed to be undirected. Since all recipients are known, recipients form a connected subgraph $G_{|R|}$ of G. The aim is to estimate the maximum likelihood (ML) estimator of with respect to the SI model.

$$\hat{s} \in \underset{s \in G_{|R|}}{\operatorname{argmax}} P(G_{|R|}|s), \tag{3.1}$$

where $P(G_{|R|}|s)$ is the probability of observing $G_{|R|}$ under the SI model. This function can be written as

$$P(G_{|R|}|s) = \sum_{\sigma \in \Omega(s, G_{|R|})} P(\sigma|s), \tag{3.2}$$

where σ represents a sequence of nodes in $G_{|R|}$ in order of the time when they get the piece of information. $\Omega(s, G_{|R|})$ is set of all such permitted propagation sequences starting with node s and consist all the nodes in the graph $G_{|R|}$. $P(\sigma|s)$ is the probability of an information propagation sequence σ with source s.

Assuming all the edges from infected nodes to uninfected nodes have equal probability, $P(\sigma|s)$ can be computed as follows. We use the tree network in Figure 3.3 for an illustration where $\sigma = (1, 2, 3, 4, 5, 6)$ and node 1 is the source. First, consider $\sigma = (1, 2)$, when the source is assumed to be 1. In that case, the next recipient node could be any of the 4 nodes: 2, 3, 7, and 8. Therefore, each one of them has a probability 1/4 to be next recipient i.e., $P(\sigma = (1, 2)|s = 1) = 1/4$. The probability of the sequence $(1, 2, 3)$ is given by $P(\sigma = (1, 2, 3)|s = 1) = P(\sigma = (1, 2, 3)|s = (1, 2)) * P(\sigma =$

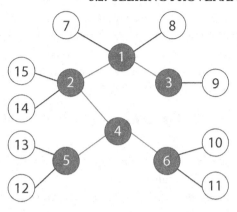

Figure 3.3: A typical tree network indicating the propagation of a given piece of information. Dark nodes are recipients (terminals).

$(1, 2)|s = 1) = 1/6 * 1/4 = 1/24$. Table 3.2 shows the step-by-step computations of $P(\sigma|s)$ for the tree network given in Figure 3.3 with $\sigma = (1, 2, 3, 4, 5, 6)$ and source $s = 1$.

Table 3.2: Computations of $P(\sigma|s)$ for the tree network given in Figure 3.3 with $\sigma = (1, 2, 3, 4, 5, 6)$ and source $s = 1$

σ	$P(\sigma|s)$
(1,2)	1/4
(1,2,3)	$1/6 * 1/4 = 1/24$
(1,2,3,4)	$1/6 * 1/24 = 1/144$
(1,2,3,4,5)	$1/7 * 1/144 = 1/1008$
(1,2,3,4,5,6)	$1/8 * 1/1008 = 1/8064$

The aim is to produce an estimate, $\hat{s}$, of the actual original source s^*, based on observation of $G_{|R|}$ and knowledge of G. Based on this setup, the maximum likelihood estimator of s^*, with respect to the SI model given $G_{|R|}$, maximizes the correct detection probability. The maximum likelihood estimator is given by

$$
\begin{aligned}
\hat{s} \quad &\in \quad \underset{s \in G_{|R|}}{\operatorname{argmax}} \, P(s|G_{|R|}) \\
&= \quad \underset{s \in G_{|R|}}{\operatorname{argmax}} \, P(G_{|R|}|s) \frac{P(s)}{P(G_{|R|})} \\
&\propto \quad \underset{s \in G_{|R|}}{\operatorname{argmax}} \, P(G_{|R|}|s) \\
&= \quad \underset{s \in G_{|R|}}{\operatorname{argmax}} \sum_{\sigma \in \Omega(s, G_{|R|})} P(\sigma|s) ,
\end{aligned}
\tag{3.3}
$$

where $P(G_{|R|}|s)$ is the probability of observing $G_{|R|}$ under the SI model, assuming s is the source s^*. $P(s)$ and $P(G_{|R|})$ are constants, under the SI model.

Let $R(s, G_{|R|})$ be the total number of distinct ways information can spread in the network $G_{|R|}$ starting from source s. For regular trees, all permitted sequences are equally likely [56]. Hence, $P(G_{|R|}|s)$ is directly proportional to $R(s, G_{|R|})$.

$$\hat{s} \quad \propto \quad \underset{s \in G_{|R|}}{\operatorname{argmax}} R(s, G_{|R|}) \cdots \text{(for regular trees)} \tag{3.4}$$

Hence, $R(s, G_{|R|})$ is referred to as *rumor centrality* of node s with respect to $G_{|R|}$. The node with the maximum rumor centrality will be called the *rumor center* or *rumor source* of the network. For trees, the rumor center turns out to be a distance center.[1] However, for general graphs, computing the rumor center is still an open problem. Equation 3.4 does not hold for general graphs, as all permitted sequences are not equally likely. Thus, the problem is computationally intensive to solve for general trees. A heuristic is proposed, based on the assumption that the recipients receive the information in a breadth-first search (BFS) fashion. Hence, Equation 3.3 becomes,

$$\hat{s} \quad \propto \quad \underset{s \in G_{|R|}}{\operatorname{argmax}} P(\sigma_s^{bfs}|s) R(s, G_{|R|}), \tag{3.5}$$

where σ_s^{bfs} is the BFS permitted sequence with node s as the source. Another issue with the general graph is that the connected subgraph $G_{|R|}$ is not known, though R is known. The proposed method approximates the $G_{|R|}$ to a BFS tree rooted at s, $T_{bfs}(s)$. Hence, Equation 3.5 becomes

$$\hat{s} \quad \propto \quad \underset{s \in G_{|R|}}{\operatorname{argmax}} P(\sigma_s^{bfs}|s) R(s, T_{bfs}(s)), \tag{3.6}$$

The above heuristic is computationally solvable and identifies a single source of a given information spread.

Identifying Multiple Sources

Lappas et al. [43] propose a method to estimate the multiple *effector* nodes, $k \geq 1$, of a given information spread with all recipients ($T = R$) known a priori, under the assumption of the IC model. Effectors are recipients such that, had the propagation started from them, it would have caused an information reception state similar to the one observed. Although effectors are not necessarily the sources of the information propagation, they are important from the point of view of observed information propagation.

This method is based on the IC model. The graph $G(V, E, p)$ is assumed to be undirected. The proposed problem necessitates finding the effector set S such that $|S| \leq k$ and utility function $U(S) = \sum_{v \in V} |a(v) - \alpha(v, S)|$ is minimized, where a denotes recipient vector, $a(v) = 1$, if $v \in V$ is a recipient, otherwise $a(v) = 0$, and $\alpha(v, S)$ denotes the probability that a node $v \in V$ receives

---

[1] A node with maximum distance centrality.

information at the end, when propagation starts from S. This problem is referred as the k-effectors problem. The k-effectors problem is NP-complete [22] under the assumption of the IC model [43]. This problem remains NP-complete even when the input graph $G(V, E, p)$ is a directed acyclic graph.

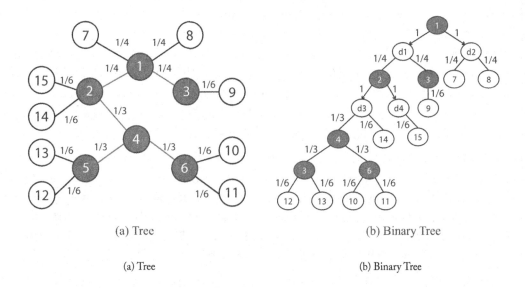

(a) Tree   (b) Binary Tree

(a) Tree   (b) Binary Tree

Figure 3.4: A typical example of conversion of a tree (a) into a binary tree (b).

The k-effectors problem can be solved optimally in polynomial time when a $G(V, E, p)$ is a tree [43]. The polynomial algorithm first converts the original tree $\mathcal{T}$ to a new binary tree $\mathcal{T}_b$, and then uses dynamic programming recursion to obtain the solution. Conversion of the original tree $\mathcal{T}$ to a new binary tree $\mathcal{T}_b$ results in the addition of many new nodes, as in Figure 3.4.

Figure 3.4 shows a typical example of conversion of a tree $\mathcal{T}$, shown in Figure 3.4(a), into a binary tree $\mathcal{T}_b$, shown in Figure 3.4(b). The process of conversion of $\mathcal{T}$ to $\mathcal{T}_b$ is as follows: Start from the root of $\mathcal{T}$, node 1, an internal node of $\mathcal{T}$ with children 2, 3, 7, and 8. Since node 1 has more than 2 children, it violates the binary tree condition of having at most 2 children per node. Node 1 is, therefore, replaced with a binary tree of depth at most log(4) and leaves 2, 3, 7, and 8. Note that such binary replacement results in the addition of some dump nodes. Let D be a set of newly added dump nodes in $\mathcal{T}_b$. Dump nodes $d1$ and $d2$ are added when the node 1 is replaced by the corresponding binary tree. As we are only allowed to select k effectors, we need to make sure that dump nodes from D are not selected. To achieve this, directed edges are added between a node that is going to be replaced with a binary tree and its newly added dump nodes, as well as between the dump nodes if they are at different levels. The direction is always from the root to the leaves, and the weight of these edges is set to 1. This transformation is repeated recursively for each child node.

The optimal solution to the k-effectors problem on T_b is same as the optimal solution of the k-effectors problem on tree T. Let $\text{OPT}(v, S, k)$ denote the cost of the best solution in the subtree rooted at node v, using at most k effectors in S. The following dynamic-programming recursion on the nodes of the tree (T_b) evaluates the optimal solution:

$$OPT(v, S, k) = \min\{OPT1, OPT2\}$$

$$OPT1 = \min_{i=0}^{k} \{OPT(r(v), S, i) + OPT(l(v), S, k - i) + U(v, S)\}$$

$$OPT2 = U(v, S \cup \{v\}) + \min_{i=0}^{k}\{OPT(r(v), S \cup \{v\}, i)$$
$$+ OPT(l(v), S \cup \{v\}, k - i - 1)\}, \tag{3.7}$$

where $r(v)$ and $l(v)$ refer to the right and left child of the node v, respectively. $U(v, S)$ refers to the contribution of node v in the utility function i.e., $U(v, S) = |a(v) - \alpha(v, S)|$ and $U(S) = \sum_{v \in V} U(v, S)$. $OPT1$ in Equation 3.7 recursion corresponds to not choosing v to be in S and $OPT2$ corresponds to choosing v to be in S. A similar check is also added so that effectors are always picked from the recipient set. We set $U(v, S) = \infty$ for every v that belongs to T_b but not to T to guarantee that actual nodes are picked as effectors. The worst case time complexity of this dynamic programming algorithm is $O(n^2 k^2)$.

For general graphs, a heuristic is proposed that aims to extract the maximum likelihood tree $\mathcal{T}$ that captures most information in G. The problem of extracting the maximum likelihood tree is NP-complete and has equivalence to the DIRECTED STEINER TREE problem [15]. The DIRECTED STEINER TREE problem asks for the directed graph $G'(V', E', w)$ with edge costs $w(u \rightarrow v) \in R^+$, a subset $T \subseteq V'$, and a root $r \in V'$, such that $\sum_{(u \rightarrow v) \in \mathcal{T}} w(u \rightarrow v)$ is minimized and subtree $\mathcal{T}$ contains directed paths from r to all the nodes in T.

Prakash et al. [53] also propose a method, called NETSLEUTH, to estimate the multiple source nodes, $k \geq 1$, of a given information spread with all recipients ($T = R$) known a priori, under the assumption of the SI model. This method determines the parameter k automatically.

3.2.2   FINDING PROVENANCE PATHS WHEN A FEW RECIPIENTS ARE KNOWN

In order to seek the provenance of information, we first aim to find the provenance paths [32]. Provenance paths are contained in a subgraph which delineates how information spreads from sources to the known terminals, including those responsible for retransmitting the information from the sources through intermediate recipients. If the provenance paths are known, the sources of information can be trivially determined. For example, roots of the provenance paths can be used as likely sources. More often than not, however, the provenance paths of information are unknown. The provenance paths problem can be formally stated as below.

Problem 3.1 The PROVENANCE-PATHS Problem   Given a directed graph $G(V, E, p)$ with known terminals $T \subseteq V$, and a positive integer constant $k \in Z^+$, find a directed subgraph G^k of G containing the provenance paths such that G^k has at the most k root nodes (sources), covers all the known terminals T, and a graph utility function, $U(G^k)$, is maximized.

$$\hat{G}^k = \underset{k \in Z^+}{\text{argmax}} \, U(G^k), \tag{3.8}$$

where $U(G^k)$ estimates utility of the provenance paths, G^k, for known terminals T. The PROVENANCE-PATHS problem aims to find $\hat{G}^k$ so as to estimate $\hat{S}$ of the original sources S^*. Utility estimation of the provenance paths $U(G^k)$ depends on the underlying information propagation model.

For a given graph G, there are exponentially many subgraphs possible having at the most k roots and covering all the known terminals T. The PROVENANCE-PATHS problem aims to extract a subgraph with the maximum utility. For any given subgraph under the IC model of information propagation, the product of all the propagation probabilities of edges estimates the likelihood of information propagation from the sources to the terminals.

$$U(G^k) = \prod_{(u \rightarrow v) \in G^k} p(u \rightarrow v) \tag{3.9}$$

The PROVENANCE-PATHS problem in this form is a *non-linear integer program*, as the objective function (Equation 3.9) is a non-linear function. The PROVENANCE-PATHS problem can be simplified into a *binary integer program* by taking logarithmic values of the input propagation probabilities.

$$\log(U(G^k)) = \log\left(\prod_{(u \rightarrow v) \in G^k} p(u \rightarrow v) \right) \tag{3.10}$$

$$= \sum_{(u \rightarrow v) \in G^k} \log(p(u \rightarrow v)) \tag{3.11}$$

The PROVENANCE-PATHS problem can also be equivalently expressed as the minimization problem by taking the negative logarithmic values of the input propagation probabilities. The minimization version of the PROVENANCE-PATHS problem is equivalent to its maximization version because the propagation probabilities are independent of any of the constraints in the original maximization version of the problem. From now on, we aim to solve the minimization version of the PROVENANCE-PATHS problem.

$$-\log(U(G^k)) = \sum_{(u \rightarrow v) \in G^k} -\log(p(u \rightarrow v)) \tag{3.12}$$

The PROVENANCE-PATHS problem is NP complete. In fact, the proposed problem is a generalization of the DIRECTED STEINER TREE problem [15]. In our context, the DIRECTED STEINER TREE problem asks for the directed graph $G(V, E, w)$ with the edge costs $w(u \rightarrow v) = -\log p(u \rightarrow v) \in \mathcal{R}^+$

(positive real number), a subset $T \subseteq V$, and a root $r \in V$, such that $\sum_{(u \to v) \in G^k} w(u \to v)$ is minimized and a subgraph G^k contains directed paths from r to all nodes of T. By the results of Zelikovsky [68], we also know that the DIRECTED STEINER TREE problem is NP-complete even for the Directed Acyclic Graphs(DAGs).

For the DIRECTED STEINER TREE problem, Charikar et al. present an approximation algorithm that achieves an approximation ratio of $O(k^\epsilon)$ in polynomial time for any $\epsilon > 0$. For $k = 1$, we can run this approximation algorithm $|V|$ times, for each node $r \in V$, and approximately solve the PROVENANCE-PATHS problem. But this solution requires knowing all the terminals a priori and can not be used to solve the multi-source problem. Also, the running time and space requirements may make this heuristic inappropriate for large-scale graphs (even for graphs with 80K+ nodes and 1.7M+ edges).

In social media, there are two major issues in approximately solving the proposed problem. First, a few terminals (less than 1% of the total recipients) are known a priori. Second, a directed graph, G, can be a large-scale graph. For example, Twitter alone consists of more than half a billion users. Hence, designing a scalable solution is a challenge. Therefore, real social networks are used to explore whether node centralities have any impact on information propagation.

In the literature, many node centrality measures have been proposed to compute the relative importance of a node in the graph. Previous research [36, 41] found that a few dominant nodes are more likely to spread the information than any random nodes. According to Wasserman and Faust [64], classical and commonly used node centralities are degree centrality, closeness centrality, betweenness centrality, and eigenvector centrality. For a large-scale network, the computation of centrality measures can be expensive, except for degree centrality [62]. For an undirected network, degree centrality of a node is determined by the number of nodes adjacent to it. Similarly, for a directed network, we have three different types of degree centralities for a node. In-degree, out-degree, and in-out-degree centralities of a node are determined by the number of nodes connected to it using incoming, outgoing, and both edges, respectively.

Our analysis indicates that, in terms of information propagation, hypotheses such as *Degree Propensity* and *Closeness Propensity* could be helpful in information propagation. Degree Propensity suggests that the higher-degree centrality nodes in a network are more likely to be transmitters than the randomly selected nodes. Closeness Propensity reveals that the higher-degree nodes closer to the terminals are more likely to be transmitters than the randomly selected higher-degree nodes. Researchers use two different ways to evaluate closeness between two nodes: hop distance, and probabilistic distance. Gundecha et al. [29] validates that the two hypotheses hold true in social networking sites.

The PROVENANCE-PATHS problem is a challenge to solve, as few terminals are known. As we have seen before, each node does not participate equally in information propagation. Based on the Degree Propensity and Closeness Propensity hypotheses, the nodes with higher-degree centralities and closer to the terminals are more likely to be transmitters. Hence, we can estimate the top m transmitters, which could have helped in information propagation, for a given set of known terminals.

Based on the knowledge of m-transmitters, the minimization version of the proposed problem can be modified as the PROVENANCE-PATH TRANSMITTERS-COVERAGE problem.

Problem 3.2 The PROVENANCE-PATH TRANSMITTERS-COVERAGE Problem   Given a directed graph $G(V, E, p)$ with known terminals $T \subseteq V$, transmitters $M \subseteq V$, and positive integer constant $k \in Z^+$, find a directed subgraph G^k of G containing the provenance paths, such that G^k has at the most k root nodes (sources), covers all the known terminals T and transmitters M, and a graph utility function $-\log(U(G^k))$ is minimized (see Equation 3.12).

Since the PROVENANCE-PATH TRANSMITTERS-COVERAGE problem is a generalized version of the PROVENANCE-PATHS problem, the PROVENANCE-PATH TRANSMITTERS-COVERAGE problem is also NP-complete and remains NP-complete even if directed graph $G(V, E, p)$ is a DAG.

Algorithm 1 PROPATHS

Input: A directed graph $G(V, E, -\log(p))$, Known Terminals $T \subseteq V$, Transmitters M, positive integers k (the number of sources).
Output: Provenance paths $G^k \subseteq G$, and Sources $S \subseteq V$.

1: $G^k \leftarrow \bigcup_{c \in M} dst(G, c, T)$
2: $S \leftarrow find\_sources(G^k)$
3: **while** $|S| \geq k$ **do**
4:     $[u, v] \leftarrow identify\_two\_closest\_nodes(G, S)$
5:     $c \leftarrow get\_shortest\_node(u, v)$
6:     $G^k \leftarrow G^k \bigcup dst(G, c, S)$
7:     $S \leftarrow find\_sources(G^k)$
8: **end while**
9: **return** $[G^k, S]$

Algorithm 1 shows the heuristic, PROPATHS, for the PROVENANCE-PATH TRANSMITTERS-COVERAGE problem. It accepts a directed graph $G(V, E, -\log(p))$ with the known terminals $T \subseteq V$, transmitters M, and a positive integer constant k (number of sources to find) as inputs. Note that we replaced the information propagation probability on each edge by its negative logarithmic value. This adjustment allows the PROPATHS to use state of the art approximation solution for the minimum DIRECTED STEINER TREE problem. The algorithm greedily computes the minimum cost solution and returns the provenance paths, $G^k \subseteq G$, and sources $S \subseteq V$ as outputs.

For each transmitter $c \in M$, we extract the minimum cost subtree rooted at c and spanning all nodes in T. Unfortunately, this subproblem is NP-hard and identical to the DIRECTED STEINER TREE problem. Although Charikar et al. [15] designed an approximation algorithm for the DIRECTED STEINER TREE problem, the running time and space requirements make this approximation inappropriate for large-scale graphs. Instead, we use a simple and efficient heuristic, $dst(G, c, T)$, for the

DIRECTED STEINER TREE problem, which combines all the shortest paths from node c to the known terminals T. The output of $dst(G, c, T)$ is a directed subgraph rooted at c. Line 1 takes the union of all such subgraphs to form a subgraph G^k. Line 2 finds all the sources from the subgraph G^k. The roots of subgraph G^k are referred to as sources. If there are no roots, we greedily decide k sources from the M terminals. Note that G^k can have at most m sources at this step.

Line 3 checks whether we have found at most k sources. If yes, we return subgraph G^k and nodes S as outputs. Otherwise, Lines 4-7 are repeated until there are at most k nodes in the set S. Line 4 identifies two nodes $u, v \in S$, which are separated by the shortest distance from a common node that has paths to nodes u and v. Line 5 finds common node $c \in V$ such that the average distance from node c to nodes u and v is the minimum. The minimum cost subtree is extracted using $dst(G, c, S)$, which is rooted at c and spans all the reachable nodes in S. This subtree is then combined with G^k at Line 6. Line 7 identifies new root nodes using the updated subgraph G^k, as described before. Finally, at line 9, the heuristic returns the provenance paths G^k and the sources S.

3.3   KEY ISSUES IN INFORMATION PROVENANCE

After we discuss a few methods that aim to address certain issues, we highlight a few key issues of future research in seeking provenance of information.

1. *Incomplete observed graph.* Solutions presented above are based on a key assumption that observed graph $G(V, E, p)$ has no missing edges or nodes. Many times, the observed graph in social media is incomplete. For instance, the different social media sites allow users to control their privacy settings. These privacy settings enable users to control their visibility to everyone on social media. Since social media is distributed and no single repository exists that stores all the interactions on social media, the assumption of a complete observed graph is far from practical. Hence, designing a robust solution to the information provenance problem that can handle an incomplete observed graph is a big challenge.

2. *Information propagation model for social media.* Researchers have been modeling the information propagation in social media for decades. Though researchers have been applying some of these, the SI, SIR, IC, and LT models, to information propagation in social media, there is still a lack of consensus on which model truly reflects the underlying propagation in social media. As pointed out before, the assumption of an underlying information propagation model can significantly change the information provenance problem setting.

3. *Network properties.* The nodes and edges are important entities of the provenance problem. Properties of nodes, such as influence and trustworthiness values, in-degree, out-degree, and different centrality measures, might facilitate in seeking the sources. Similarly, properties of edges, such as time lag between connected nodes, cross media edges, etc., help guide the search for information provenance. Information propagation in a social media environment exhibits certain properties based on the type of origin. For example, a popular source leads to a shallow

cascade compared to that of a non-influential source. Also, sentiments become polarized as the depth of the cascade increases [46]. Hence, exploring network properties can provide clues to the information provenance problem.

4. *Speedily determining the sources.* Social media is very dynamic. In some cases, the value of information can depreciate quickly. For example, all the information attributes related to the 2011 Japan earthquake may not be important after a certain time period. Hence, in certain scenarios speedily identifying the sources is critical.

5. *Reducing the search space.* The social media graph can be large. Methods based on the provenance paths problem need to traverse many search directions. Hence, reducing the search space is essential from the scalability point of view. One possible way to reduce the search space is to estimate node credibility in propagating information. For example, social news media like CNN and BBC have greater credibility value than random users in Facebook. Friends are more credible than random strangers. In the next chapter, we discuss how provenance attributes can be used to reduce the search space.

3.4   SUMMARY

In this chapter, we first presented the information propagation models commonly assumed in seeking provenance of information. Our focus was to exploit the social media network structure alone to get closer to the actual sources. Then we discussed two approaches to finding sources and solving the PROVENANCE-PATHS problem. For each approach, we described a few representative solutions in detail. A few key issues are also highlighted to discuss the possible future research directions in seeking provenance of information

All the methods discussed above are based on the observed social media network $G(V, E, p)$. This can be a large-scale network. For example, Facebook alone consists of billions of users with many edges. Hence, designing a scalable solution is a challenge. We address this issue in the next chapter with the help of provenance attributes.

CHAPTER 4

Provenance Data

In Chapter 2, we analyzed provenance attributes of a given node in social media. In Chapter 3, we discussed different methods of seeking provenance of information using network information alone. In this chapter, we introduce a framework for provenance data search using both provenance-related information and network information to limit the search space while dealing with large social media networks. Provenance data consists of sources of information, provenance paths from sources to terminals, and provenance-related information, which includes attribute values and propagation history of all nodes along these paths. Propagation history provides the recent information spread from a certain node and gives a likelihood prediction for future propagation. Searching for provenance data has two objectives: first, to characterize the sources of given information, and second, to uncover provenance paths as accurately as possible.

In this chapter, we first describe the framework for provenance data search and show how the algorithm employed in this framework aims to find provenance data. Then we show additional information: propagation history can be used to help provenance data search. Using a real disinformation spread example in social media about the Assam Exodus, we show that the framework can be applied to a cross-platform network and help to identify the likely rumormongers or disinformation centers.

4.1   AN ITERATIVE FRAMEWORK

The framework for provenance data search is based on an iterative method, where every iteration makes use of provenance-related information to reduce the search space and guide the provenance search. As shown in Figure 4.1, an iteration for provenance data search consists of five steps: (1) extracting a local network, (2) collecting provenance-related information, (3) updating the local network, (4) recovering provenance paths, and (5) identifying sources. Next, we discuss the framework for provenance data search by expatiating these five steps.

1. *Extracting a local network:* The main task is to extract the m-hop local directed graphs for the sources estimated so far S, $G_S = \{G_s | s \in S\}$, where positive integer $m \geq 1$, and G_s represents the m-hop directed graph from the incoming links of $s \in S$. On every iteration, the framework makes a new estimation of sources S. They are initialized to terminals T at the start. Each G_s consists of all the nodes and edges that can reach node s in less than or equal to m hops.

2. *Collecting provenance attributes and propagation history:* At this step, we aim to collect provenance attributes and propagation history for all the nodes in G_S. The cost of collecting is

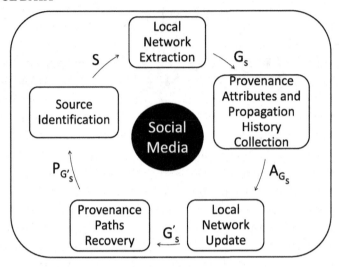

Figure 4.1: Framework for provenance data search.

directly proportional to the number of nodes in G_S. In Appendix B, we discuss the tool of collecting provenance attributes.

3. *Updating local networks:* Based on the provenance attributes, the local graphs G_S can be updated. New propagation edges are added for nodes, based on their list of potential receivers. Also, the information propagation probability for a given edge $(u \rightarrow v) \in G_S$ is estimated based on the similarity scores of collected provenance attributes of nodes u and v. For example, in the case of the Assam Exodus, an Indian Twitter user is more likely to propagate information to his other Indian Twitter followers than random Twitter users. If provenance attributes are not available for any node $u \in V$, we assume that the information propagation probability on every outgoing edge from node u to be the same and equal to the reciprocal of the total number of outgoing edges. Also, propagation history can provide additional information to update local networks. In next section, we present the methodology of propagation history.

4. *Recovering provenance paths:* Here, we aim to recover the provenance paths (subgraph) using updated local directed graphs G'_S. The recovered paths $P_{G'_S}$ represent how the information most likely traveled. In Chapter 3 we show that it is feasible to recover provenance paths [43, 53, 56].

5. *Identifying sources:* At this step, nodes without in-degrees in subgraph $P_{G'_S}$ are identified as potential sources.

Following the five steps of the framework, we design the algorithm for provenance data search. Algorithm 2 shows the process for provenance data searching. It accepts the set of terminal nodes

---

Algorithm 2 SearchProvenanceData

Input: Terminal set $T \subseteq V$, positive integers k (sources) and m (hops).
Output: Source set $S \subseteq V$, provenance paths P_{G_S} and attribute values and propagation history A_{G_S}

 $S \leftarrow T$
 repeat
  1: $G_S \leftarrow local\_network\_extraction(m, S)$
  2: $A_{G_S} \leftarrow provenance\_attributes\_and\_propagation\_history(V(G_S))$
  3: $G'_S \leftarrow local\_network\_update(A_{G_S})$
  4: $P_{G'_S} \leftarrow provenance\_paths(G'_S, S)$
  5: $S \leftarrow source\_nodes(P_{G'_S})$
 until $|S| \leq k$
 $A_{G_S} \leftarrow provenance\_attributes\_and\_propagation\_history(V(G_S))$
 return S, P_{G_S}, A_{G_S}

$T \subseteq V$ and positive integer constants k (number of sources to find) and m (hop length for local graphs) as input and returns the source set $S \subseteq V$ as output. Sources, S, are initialized to terminals, T. The first step extracts m-hop directed graphs G_S. Step 2 collects A_{G_S}, provenance attributes for all the nodes in G_S. Provenance attributes and propagation history information A_{G_S} are used to update G_S, resulting in G'_S in step 3. Step 4 recovers provenance paths, $P_{G'_S}$, using updated G'_S and identified sources S. Nodes with no in-degrees in $P_{G'_S}$ are identified as new sources S at step 5. Provenance paths, $P_{G'_S}$, represent how the information has most likely traveled from sources S in graphs G'_S. The 5 steps are repeated until at most k sources are not found. Finally, the algorithm returns the identified sources S, provenance paths P_{G_S}, and provenance attributes values and propagation history information A_{G_S} of the nodes along the paths. The algorithm does not need to know the entire network at the start. On every iteration it only requires m-hop directed graphs from the identified sources S.

  In the following section, we will discuss the mechanism of propagation history regarding provenance data search.

4.2 PROPAGATION HISTORY

In Chapter 2, we describe how to collect provenance attributes that help to limit the search space. Besides provenance attributes, determining propagation history [63] provides additional information to guide the provenance data search. Propagation history can be used when the provenance attributes are not available. Nodes that spread information in the past are more likely to be on the path for the current event. We take Twitter as an example to illustrate the process in terms of features extraction, ranking measures, and follower ranking prediction methods.

4.2.1 FEATURE EXTRACTION

The Twitter social network can be presented as a directed graph $G = \{V, E\}$, where $V = \{u_1, u_2, \ldots, u_n\}$ is the set of users and E is the following relationship between users. A typical Twitter user u has a set of followers ($Follower(u)$) and friends ($Friend(u)$) which are also known as followees. We denote contacts ($Contact(u)$) as the union of the user's followers and friends; that is,

$$Contact(u) = Follower(u) \cup Friend(u) . \tag{4.1}$$

Friends, followers, and contacts are *neighbors* of a user, as they are connected in a certain manner. The cardinality of a set represents its size, e.g., $|Friend(u)|$ represents the number of friends of user u.

Common friends CFR refer to the set of users who are followed by two users u_i and u_j. Similarly, we define the common followers CFO and common contacts CCO as the users who are shared by the two corresponding sets, i.e.,

$$
\begin{aligned}
CFR(u_i, u_j) &= Friend(u_i) \cap Friend(u_j), \\
CFO(u_i, u_j) &= Follower(u_i) \cap Follower(u_j), \text{ and} \\
CCO(u_i, u_j) &= Contact(u_i) \cap Contact(u_j) .
\end{aligned}
\tag{4.2}
$$

We aggregate all tweets that are owned by user u, then form a term-frequency vector $t(u)$, excluding stop words. Similarly, the set of hashtags and URLs that are associated to user u are represented as term-frequency vectors $ht(u)$ and $url(u)$, respectively.

Given a user u and her followers, our primary focus is to rank the followers by their likelihood of retweeting any one of her tweets, considering a wide range of features from the Twitter social network and user-generated content. The top-k followers most likely to retweet are returned as active retweeters of this user. Let $P(f_i|u)$ be the retweet likelihood of the i-th follower from u, the objective function of identifying active retweeters is defined as follows,

$$\max_{\{f_i\}_{i=1}^{k}} \quad \sum_{i=1}^{k} P(f_i|u) \tag{4.3}$$

$$s.t. \quad f_i \in Follower(u) .$$

We automatically rank a user's followers by their likelihood of future retweeting. Our hypothesis is that the extent to which a person may retweet from her friends can be learned from her online behaviors, interactions, etc. Boyd et al. [10] summarized several reasons why people retweet from their friends. For example, where a tweet is informative, the followers want to share it with their own followers or save it for future personal access, make a stance as agreeing with someone, show support and presence as a listener, start a conversation, etc.

We extract some features that may contribute to conduct the follower ranking. These features include user similarity, online interaction, structural features, and user profiles. Some features

Table 4.1: Description of features to rank followers

Group	Feature	Description
Proximity	Common Followers	Number of users who follow both users
	Common Friends	Number of users who are followed by both users
	Common Contacts	Number of users who have connection with both users
	Mutual Links	Indicator of whether two users follow each other
	Social Status	Larger PageRank values represent higher social status, and vice versa
Content	Common Hashtags	Number of common hashtags that are used by both users
	Common URLs	Number of common URLs that are shown in both users' tweets
	Tweet Similarity	Cosine similarity between the two users' tweet vector $t(u)$
Interaction	Reply	Number of replies from one user to another
	Mention	Number of times that one user mentions the other in his or her tweets
Profile	Status	Number of Tweets by user
	Lists	Number of lists belonging to a user
	Language	Preferred language of a user
	Account	Account creating date
	Friends	Number of a user's friends
	Followers	Number of a user's followers
	Contacts	Number of a user's contacts

are well discussed in prior work such as [48, 52, 61, 67]. Table 4.1 lists all features that can be roughly categorized into four groups by their functions: proximity, content, interaction, and profile. **Proximity-based features** measure the similarity between an arbitrary pair of following users u_i and u_j, relative to the network topology. These features are extracted from the Twitter following network, and therefore, are irrelevant to retweeting content. Features include common friends, common followers, common contacts, social status, etc. **Content-based features** measure the similarity of the user-generated content between two users. The set of features used in this paper are common hashtags, common URLs, and tweet similarity. **Interaction-based features** indicate the frequency with which two persons talk to each other. We extract the number of replies and mentions between a pair of users as the interaction features. **Profile-based features** include the statistics related to each user: the status (or tweets), friends, followers, contact counts, list count, the language a person uses, and the account creation date.

Usually a retweet is characterized by the abbreviation "RT" at the beginning of the tweet. The "@" sign followed by a user name indicates that it is a mention or reply to the user. We consider retweet as information diffusion, while mention and reply are considered as interactions between users. A hashtag is used to group posts by their topics; e.g., a tweet containing hashtag "#egypt" implies that it may be related to Egypt. A hashtag could be any word or phrase that is prefixed with a "#" sign. Also, many tweets are embedded with URLs.

4.2.2 RANKING MEASURES

The set of approaches that are potentially suitable for ranking a user's followers by their likelihood of retweeting are summarized. All these methods assign a retweeting score to an arbitrary pair of following relationships, i.e., $P(f_i|u) \in [0, 1]$, $f_i \in Follower(u)$. Some methods are very well developed, but are applicable in other tasks. To simplify notations and for ease of understanding, we use the hashtags as an example to derive the proposed approaches. The definitions can be generalized to other features easily. Assume u_i and u_j are two Twitter users that have a following relationship; e.g., u_i is a follower of u_j. The ranking measures are listed next.

- **Shared feature counting.** Countable features in this data set include shared neighbors (i.e., friends, followers, and contacts), shared hashtags, and shared URLs. This approach is reasonable because shared features and retweet likelihood are correlated.

$$|ht(u_i) \cap ht(u_j)| \tag{4.4}$$

- **Jaccard Index** measures the extent to which two sets overlap. It is a normalized similarity measure and its value is between 0 and 1.

$$\frac{|ht(u_i) \cap ht(u_j)|}{|ht(u_i) \cup ht(u_j)|} \tag{4.5}$$

- **Adamic/Adar Index** assigns more weight to shared features that are rarely used by other people [1]. We consider the hashtags and URLs that are used by Twitter users in the paper to compute this index. Let u_i and u_j be two users, z be a shared hashtag, and $F(z)$ represent the number of users who used the feature z. The Adamic/Adar index between two users is given by

$$\sum_{z \in ht(u_i) \cap ht(u_j)} \frac{1}{\log F(z)} . \tag{4.6}$$

We also consider a variation (i.e., Weighted Adamic/Adar Index) where the number of times that a hashtag has been shared by two users is used. Let z_{u_i} be the number times that a hashtag z is used by user u_i. The definition is shown as follows:

$$\sum_{z \in ht(u_i) \cap ht(u_j)} \frac{\min(z_{u_i}, z_{u_j})}{\log F(z)} . \tag{4.7}$$

- **Tweet similarity** is computed by assuming each user as a term-frequency vector after removing stop words. The tweet similarity between two users u_i and u_j is given by the vector similarity,

$$\frac{t(u_i) \cdot t(u_j)}{\|t(u_i)\| \cdot \|t(u_j)\|} . \tag{4.8}$$

4.2.3   FOLLOWER-RANKING PREDICTION METHODS

The followers, friends, and contacts of a user are all deemed as neighbors. We examine all following pairs in the Twitter social network and find that the retweet likelihood increases as the number of common neighbors increases. However, the common neighbors are not strong indicators of retweeting likelihood, as we notice that the percentages are less than 3%, even two users share as many as 100 neighbors.

Hashtags that are shared by a pair of users are correlated to the times of retweeting. Two variations of common hashtag computation strategies are used in the experiments: weighted and unweighted. The unweighted variation is exactly computed by Equation (4.4), while the weighted version is slightly different, by taking the shared frequency into account. Its definition is given below,

$$\sum_{z \in ht(u_i) \cap ht(u_j)} \min(z_{u_i}, z_{u_j}), \qquad (4.9)$$

where z_{u_i} and z_{u_j} represent the number of times that a hashtag z is used by users u_i and u_j, respectively. Our hypothesis is that two users who use the same set of tags more frequently are more likely to retweet from each other.

URLs in tweets are mostly references to external sources where the tweet is inspired. Similar to hashtags, we consider the shared URLs by two different strategies: weighted and unweighted. The measures for common URLs are transplantable from the definitions of common hashtags. The retweet likelihood is positively correlated to the number of shared URLs between two users. The retweet probability increases quickly when only few URLs are shared, but then the trend becomes flat as more URLs are shared.

Given a pair of users u_i and u_j, the tweet similarity is defined as the cosine similarity of their tweet vectors $t(u_i)$ and $t(u_j)$. Intuitively, two users with a higher tweet similarity are more likely to share certain interests, thus increasing the likelihood of retweeting.

Reply and mention are two types of interactions that are used to predict retweeting. As shown in the two figures, the trend of both interactions are similar: both of them increase significantly if two users have few interactions, then become flat. In addition, the trend of replies (Figure 4.2(a)) shows larger variance than that in mentions (Figure 4.2(b)). Although both reply and mention are strong indicators for retweeting, the ratios of replies and mentions to the total number of tweets are small, which limits the effectiveness in retweet prediction.

We evaluate the performance of the above methods by their top-k precision comparing to the actual retweet history. Top-k precision is widely used in information retrieval tasks. Specifically, for each user, we rank the followers by their likelihood to retweet from the user in descending order, then compare the top-k ranked users. The number k is chosen as 1, 5, 10, 20, 30, 40, 50, 100, and 500. The precision averaged over all users in the Twitter social network is reported.

Table 4.2 lists the precision performance of the different methods. Each column represents the top-k users that are retrieved, e.g., column 1 indicates that we only consider the first user who is recommended by the corresponding methods.

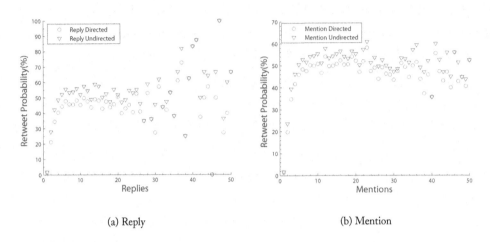

(a) Reply   (b) Mention

Figure 4.2: Retweet probability vs. number of interactions. (Communication direction is considered).

Table 4.2: Precision performance of various methods

	Method	Top k Retrieved Followers								
		1	5	10	20	30	40	50	100	500
Hashtag	Common Tags	.29	.18	.15	.13	.11	.11	.10	.09	.07
	Jaccard Index	.26	.16	.13	.11	.10	.10	.09	.08	.07
	Adamic/Adar	.33	.20	.16	.13	.12	.11	.10	.09	.07
	Weighted Adamic/Adar	.29	.18	.15	.12	.11	.11	.10	.09	.07
URL	Common URLs	.42	.25	.19	.15	.13	.12	.11	.09	.07
	Jaccard Index	.41	.24	.18	.14	.12	.11	.10	.09	.07
	Adamic/Adar	**.47**	**.28**	**.21**	**.16**	**.14**	.12	.11	.09	.07
	Weighted Adamic/Adar	.47	.28	.21	.16	.13	.12	.11	.09	.07
Neighbor	Common Friends	.09	.07	.07	.07	.07	.07	.07	.06	.06
	Jaccard Index (CFR)	.15	.10	.09	.08	.07	.07	.07	.07	.06
	Common Followers	.11	.09	.08	.08	.08	.07	.07	.07	.06
	Jaccard Index (CFO)	.15	.11	.10	.09	.08	.08	.08	.07	.06
	Common Contacts	.10	.08	.08	.07	.07	.07	.07	.07	.06
	Jaccard Index (CCO)	.16	.11	.09	.08	.08	.07	.07	.07	.06
Interaction	Reply	.15	.13	.13	.12	.12	.12	.12	.12	.12
	Mention	.18	.15	.14	.14	.14	**.14**	**.14**	**.13**	**.13**
Similarity	Tweet	.37	.21	.16	.13	.12	.11	.11	.10	.08

Methods based on URLs work best. In most retrieval or recommendation applications, k is typically chosen to be a small number, e.g., 10. The URL-based methods outperform the other methods with a margin, especially when the selected number k is small, e.g., the best performance of the URL-based approach is 11.9% better than the second best approach when $k = 1$. We also notice that different features have different strengths in retweet prediction: URL is the best, followed by tweet similarity, hashtag, interaction, and common neighbors. Statistically, comparing the best performances of URL-based methods to those of feature-based methods, the relative improvements are 30.5%, 72.4%, 99.3%, and 159.3%, respectively. This result is consistent with prior studies that tweets with URLs are more likely to be retweeted by others [42, 48, 61].

4.3   A CASE STUDY OF DISINFORMATION IN ASSAM EXODUS

In this section, we use the illustrative example of the "Assam Exodus" to demonstrate how the proposed framework can facilitate identification of rumormongers or disinformation centers in social media.

Extracted real disinformation spread. A gory video clip[1] of riots in Indonesia was posted on YouTube, labeled as a video of the Assam riots. Figure 4.3 shows the partial spread of this video clip ($y1$) in social media. YouTube later removed the original video clip due to its shocking and disgusting content. The majority of tweets containing a link to this video clip were spread on the same day (August 16, 2012) in Twitter. Although some recipients later removed these tweets, as of October 2012, many of them are still accessible.

The information spread shown in Figure 4.3 is generated using Twitter and Topsy.[2] Each node represents an entity. An entity refers to either a Twitter user (node label starts with "t"), a YouTube video link (node label starts with "y"), or a web article link (node label starts with "w"). Directed links show direction of information flow.

Figure 4.3 shows a few Twitter users, including $t1, t5, t13, t15, t17, t22, t50, t51$, and $t61$, tweeted about the video clip $y1$. When a user tweets on Twitter, the tweet is visible to all his followers. For simplicity, we remove some outgoing edges from users to their followers, unless followers have re-tweeted further. A few Twitter users (e.g., $t1$ and $t67$) are also found to propagate the tweet to other non-follower Twitter users, especially influentials, to maximize propagation. Figure 4.3 shows explicit information propagation edges. Most of these influentials have considerably large numbers of followers. The influentials include Indian politicians,[3] journalists, TV news reporters, and social activists. Some Twitter followers of $t67, t72$ and $t75$ re-tweeted. 16, 6, and 3 Twitter followers (seen in blue circles, where the size of a circle is proportional to the number of followers) of users $t67, t75$, and $t72$, respectively, retweeted further. We used the Youtube video link to collect the nodes in part Y, and used a news article link which is found in part A to collect the nodes carrying the news link (Part A).

Characteristics of different propagation paths. Users A, B, and C receive tweets from twitter users $t4, t42$, and $t57$, respectively. Figure 4.3 highlights propagation paths of a video from source node $y1$ to users A, B, and C. Table 4.3 shows the characteristics of these paths. Starting from YouTube, path 1 traverses multiple platforms before reaching the node A, whereas other paths,

---

[1]http://www.youtube.com/watch?&v=4gdpAP3WkH4, accessed in Oct 2012.
[2]http://www.topsy.com/
[3]A few of them are even incumbent ministers at the state (Assam) or national (India) level.

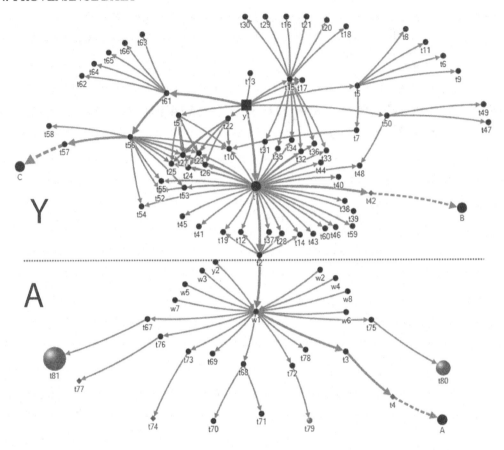

Figure 4.3: An example of disinformation spread in social media about the Assam Exodus.

2 and 3, traverse only in Twitter before reaching the users B and C, respectively. Paths 1 and 2 spread through the node $t1$, who plays a major role in disinformation spread, and traverse 6 and 3 hops, respectively. Path 3 spreads information using Twitter users who were not spreading the disinformation explicitly, and traverses 4 hops before reaching the user C.

Searching for provenance data. We apply the algorithm proposed in Section 4 to search for provenance data from the perspective of querying users A, B, and C. Table 4.4 shows experimental results for single-source identification based on 1-hop and 2-hops directed graphs. The Twitter graph obtained based on the 2-hops network from each node in Figure 4.3 consists of millions of nodes. Previous network based approaches [43, 53, 56] are computationally intensive and cannot be applied to locate sources that require knowledge of all the recipients.

User A receives the tweet from terminal $t4$. The 1-hop local graph of $t4$ consists of at least 148 potential nodes (whom he is following). We collect available provenance attributes of all these nodes.

Table 4.3: Propagation paths from source node $y1$ to nodes A, B, and C, respectively

ID	Path	Hops
1	$y1 \rightarrow t1 \rightarrow t2 \rightarrow w1 \rightarrow t3 \rightarrow t4 \rightarrow A$	6
2	$y1 \rightarrow t1 \rightarrow t42 \rightarrow B$	3
3	$y1 \rightarrow t61 \rightarrow t56 \rightarrow t57 \rightarrow C$	4

Table 4.4: Experimental results for single-source identification based on 1-hop and 2-hops directed graphs

		$k = 1$	
Users	Terminals	$m = 1$	$m = 2$
A	$\{t4\}$	$t3$	$w1$
B	$\{t42\}$	$t1$	$y1$
C	$\{t57\}$	$t56$	$t61$
$\{A, C\}$	$\{t4, t57\}$	$y1$	–

Based on provenance attributes of $t4$, propagation probability for an edge $(t3 \rightarrow t4)$ is recovered as 1, since $t4$ retweets tweets from $t3$. Hence, the 1-hop directed graph extracted at $t4$ identifies $t3$ as a source. The 2-hops directed graph extracted at $t4$ consists of more than 200k nodes. We collect provenance attributes of all these nodes. Based on provenance attributes of $t3$, propagation probability for an edge $(t3 \rightarrow w1)$ is very high (above 0.9), since link $w1$ is mentioned in $t3$. Also, we find that a news website, whose Twitter handle[4] is $t3$, is hosting $w1$. Hence, the 2-hops directed graph extracted at $t4$ identifies $w1$ as a source (with 90% confidence). This shows that source identification gets more accurate ($w1$, instead of $t3$) as the size of extracted directed graphs increases. But this improvement comes at the cost of more time spent in collecting provenance attributes, and recovering provenance paths. We can make a better source estimation for user A by changing the convergence condition in the proposed algorithm, such that sources are at least h-hops (a positive integer constant) away from terminals.

For user B, the Provenance Data Framework proposed in Section 4.1, accurately identifies the video clip $y1$ as a source. This is primarily because terminal $t42$ receives the tweet from user $t1$, who played a vital role in spreading video clip $y1$. As of October 2012, we can still access much of the information spread from user $t1$ using Topsy search. We also conducted the experiment for multiple known terminals. If $t57$ and $t4$ are known terminals due to users A and C, then the algorithm with iterative 1-hop directed graphs extracted at intermediary sources converges at node $y1$ and identifies it as a source. This shows the algorithm makes better estimations, if more terminals are known at the start. The algorithm does not need to know the entire network at the start. Thus, it can handle large cross-platform networks.

---

[4]Twitter username

4.4 SUMMARY

In this chapter, we first presented the framework for provenance data search using both provenance-related information and network information. Then we showed that, besides provenance attributes, the propagation history can provide additional information in this framework. We gave a real-world case study of disinformation spread in Assam Exodus to show how the framework works in identifying the rumormonger.

The INFORMATION PROVENANCE problem answers which nodes are the possible sources of some particular information, say a text message or a tweet. We present some key research issues in this burgeoning area below.

1. What are the characteristics of sources such that we can identify a source when we encounter one? That is a challenging task because source nodes are not necessarily without incoming links in social media networks.

2. How can we use different types of social media data for provenance data search? Content, user profiles, propagation history, and interaction patterns can play complementary roles in backtracking information propagation. As a popular source can lead to a shallow cascade [44], the study of node centrality measures can be of great help.

3. How can we infer missing links in searching for provenance data with partial information? By the nature of social media, most information is informal and partial. Links can expand the network (i.e., new nodes are added), and data associated with a node provides more information, though still partial.

4. How can we limit the search space in the vast land of social media? It is incumbent on us to develop a scalable solution for provenance data search.

5. What are effective and objective ways of verifying and comparing different approaches to the information provenance problem? Lack of ground truth constitutes the foremost difficulty.

6. Entity resolution [19] refers to the task of distinguishing whether pieces of information belong to a propagation. It is important to find the recipients and terminals for a piece of information.

7. Information propagation crosses multiple social media sites [66]. How to find the connections among them are still an open challenge.

The INFORMATION PROVENANCE problem is an unprecedented challenge. The abundance of data in social media provides ways to tackle the problem. In this lecture we present the preliminary findings of studying the INFORMATION PROVENANCE problem, and its research progress can pave the way for many equally challenging and important issues, such as source trustworthiness, information reliability, and user credibility.

APPENDIX B

Online Provenance Data Tool

Provenance data collector[1] is an online data collection tool focusing on efficiently retrieving useful attribute values of a given twitter user. This tool features an intuitive user interface and is designed to enable fast retrieval of a maximum number of desired provenance attributes [31]. If some desired provenance attributes are uncertain, the tool provides the best possible URL (Uniform Resource Locator) to help users find them further. In addition to provenance attributes, the tool also presents other attribute values and related images during the search and measures to evaluate efficiency of the system. Figure B.1, shows an overview of the tool for collecting provenance attribute values. Next, we give a detailed description of the tool consisting of three major components.

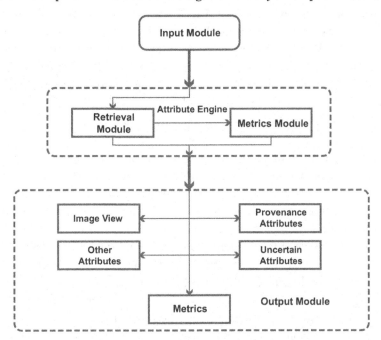

Figure B.1: Overview of the tool for collecting provenance attribute values.

The input module asks social media users to perform two tasks: input the twitter identifier and, select attributes of interest from a list of attributes. We use the twitter handle to uniquely

---

[1]The provenance data collector tool is located at http://blogtrackers.fulton.asu.edu/Prov_Attr, and demonstration video can be found at http://www.youtube.com/watch?v=W4hbhyVu6zw.

identify each twitter user. Each twitter handle is prefixed by @. For example, the unique twitter handle for President Barack Obama is "@barackobama." User selections of attributes are referred to as *provenance attributes.* The attribute engine then uses the twitter handle and provenance attributes to retrieve useful information.

The attribute engine is at the core of the provenance data collector tool. The primary objective is to retrieve values of provenance attributes, and compute measures to evaluate the efficiency of the tool.

The attribute retrieval module takes a twitter handle and provenance attributes from the input module and explores different social media sites for information collection. The attribute retrieval module utilizes four main important social media sites for mining the attribute values: Twitter profile, LinkedIn public profile, Wikipedia page, and search engines results from Google and Bing. Using the twitter handle, formal name and location can be obtained from the Twitter user's profile. Both attribute values are then queried on different search engines, including Google and Bing. The user profile from professional social networking sites, such as LinkedIn, is retrieved from the search results. The publicly available LinkedIn profile page is then mined for the attributes. LinkedIn consists of 200+ million users.[2] We find that, if a twitter user is also available in LinkedIn, values of many attributes can be collected from their publicly available profile page. Some (popular) users also have their own Wikipedia pages, which are also used by the module to obtain attribute values. Attributes obtained from the above sites use different information retrieval techniques.

In addition to provenance attributes, we also keep records of other collectible attribute values from each visited social media site. To provide further authentication for the collected attribute values, we also retrieve related images of an input twitter user, using image results from Google and Bing search engines.

Based on all the attribute values collected by the retrieval module, the metrics module computes three measures: information availability, information legitimacy, and retrieval speed. These measures help us to evaluate the efficiency of our system as well as provide a way to compare and contrast the information about different twitter users. Information availability and information legitimacy are described in Chapter 2.

The output module obtains the attribute values and the metrics, segregates them into categories, and presents in easily readable formats. The output module is segregated into five sections; four sections corresponding to different attribute categories and one presenting the provenance metrics. In Figure B.2, we see the web interface for the provenance data collector tool, showing provenance attribute values of President Barack Obama (@barackobama).

The upper left section (images) shows the images related to the twitter user. Visual information plays a significant role in shaping user confidence about the values of collected provenance attributes. The upper right section (provenance attributes) displays the values of those provenance attributes that can be found using our retrieval model. The number of social media sites from which the particular value is verified, along with the URLs to the site, are presented alongside each provenance attribute.

---

[2]http://en.wikipedia.org/wiki/LinkedIn, accessed on Dec 1, 2012.

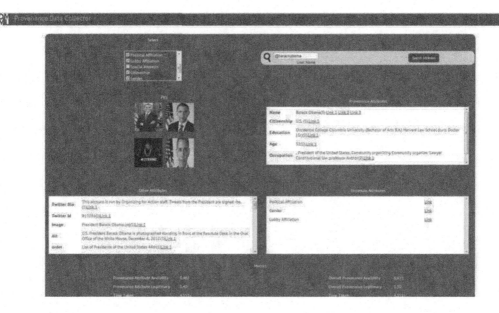

Figure B.2: Web interface of the Provenance Data Collector tool showing provenance attribute values of President Barack Obama (@barackobama).

The lower left section (other attributes) shows additional attributes retrieved from different sites. These attributes, although not specifically asked for by the user, present diverse viewpoints about an input twitter user. Provenance attributes with uncertain values are presented in the lower right section (uncertain attributes). In this case, the user is directed to the most relevant URL where she might be able to find more information. Information available in this section is an indication of information which is hard to retrieve. The bottom section displays the provenance metrics.

Bibliography

[1] L. A. Adamic and E. Adar. Friends and neighbors on the web. *Social Networks*, 25(3):211–230, 2003. DOI: 10.1016/S0378-8733(03)00009-1. 52

[2] R. Adams. John Roberts retirement rumour: A lesson in gossip and the internet. *The Guardian*, March 5, 2010. 8

[3] S. Ahsan and A. Shah. *Designing Software-Intensive Systems : Methods and Principles*, chapter Quality Metrics for Evaluating Data Provenance, pages 455–473. Information Science Reference (an imprint of IGI Global), 701 E. Chocalate Ave, Suite 200, Hershey, PA 17033, 2009. 21, 22, 23

[4] R. Allain. Tweet waves vs. seismic waves. *Wired Science Blogs*, August 26 2011. http://www.wired.com/wiredscience/2011/08/tweetwaves-vs-seismic-waves/ referenced March 6, 2011. 8

[5] M. K. Anand, S. Bowers, T. McPhillips, and B. Ludäscher. Efficient provenance storage over nested data collections. In *EDBT '09: Proceedings of the 12th International Conference on Extending Database Technology*, pages 958–969, New York, NY, USA, 2009. ACM. DOI: 10.1145/1516360.1516470. 22

[6] R. Anderson, R. May, and B. Anderson. *Infectious Diseases of Humans: Dynamics and Control*, volume 28. Wiley Online Library, 1992. 31, 32

[7] G. Barbier and H. Liu. Information Provenance in Social Media. In J. Salerno, S. J. Yang, D. Nau, and S.-K. Chai, editors, *The 4th International Conference on Social Computing, Behavioral Modeling, and Prediction (SBP)*, volume 6589 of *Lecture Notes in Computer Science*, pages 276–283, College Park, MD, March 2011. Springer. DOI: 10.1007/978-3-642-19656-0. 13

[8] G. P. Barbier. Finding provenance data in social media. Dissertation, Arizona State University, Tempe, AZ, December 2011. 14, 17, 19, 20, 22, 23, 24, 25, 27, 28, 29, 30

[9] M. Block. Tracing rumor of John Roberts' retirement. National Public Radio, March 2010. accessed on October 19, 2011. 10

[10] D. Boyd, S. Golder, and G. Lotan. Tweet, tweet, retweet: Conversational aspects of retweeting on twitter. In *Proceedings of the 43rd Hawaii International Conference on Social Systems*, pages 1–10, 2010. DOI: 10.1109/HICSS.2010.412. 50

[11] N. M. Bradburn, S. Sudman, and B. Wansink. *Asking Questions*. John Wiley & Sons Inc., 2004. 17

[12] P. Buneman, S. Khanna, and T. Wang-Chiew. Why and where: A characterization of data provenance. In J. Van den Bussche and V. Vianu, editors, *Database Theory — ICDT 2001*, volume 1973 of *Lecture Notes in Computer Science*, pages 316–330. Springer Berlin / Heidelberg, 2001. 10.1007/3-540-44503-X20. DOI: 10.1007/3-540-44503-X. 13

[13] D. Chakrabarti, Y. Wang, C. Wang, J. Leskovec, and C. Faloutsos. Epidemic Thresholds in Real Networks. *ACM Transactions on Information and System Security (TISSEC)*, 10(4):1, 2008. DOI: 10.1145/1284680.1284681. 31, 35

[14] A. Chapman and H. Jagadish. Issues in building practical provenance systems. *Bulletin of the IEEE Computer Society Technical Committee on Data Engineering*, 40(4):38–43, 2007. 13

[15] M. Charikar, C. Chekuri, T. Cheung, Z. Dai, A. Goel, S. Guha, and M. Li. Approximation Algorithms for Directed Steiner Problems. In *Proceedings of the 9th annual ACM-SIAM symposium on Discrete algorithms*, pages 192–200, 1998. DOI: 10.1006/jagm.1999.1042. 40, 41, 43

[16] S. Coppens, D. Garijo, J. M. Gomez, P. Missier, J. Myers, S. Sahoo, and J. Zhao. Provenance XG Final Report. Final Report XGR-prov-20101214, World Wide Web Consortium (W3C), December 2010. 12, 13

[17] B. Corcoran, N. Swamy, and M. Hicks. Combining provenance and security policies in a web-based document management system. In *On-line Proceedings of the Workshop on Principles of Provenance (PrOPr)*, Nov. 2007. `http://homepages.inf.ed.ac.uk/jcheney/propr/`, accessed on October 19, 2011. 23

[18] D. Easley and J. Kleinberg. *Networks, crowds, and markets: Reasoning about a highly connected world*. Cambridge Univ Pr, 2010. DOI: 10.1017/CBO9780511761942. 31

[19] A. Elmagarmid, P. Ipeirotis, and V. Verykios. Duplicate record detection: A survey. *Knowledge and Data Engineering, IEEE Transactions on*, 19(1):1–16, Jan. DOI: 10.1109/TKDE.2007.250581. 58

[20] H. Gao, G. Barbier, and R. Goolsby. Harnessing the crowdsourcing power of social media for disaster relief. *Intelligent Systems, IEEE*, 26(3):10 –14, may-june 2011. DOI: 10.1109/MIS.2011.52. 3, 8

[21] H. Gao, X. Wang, G. Barbier, and H. Liu. Promoting coordination for disaster relief - from crowdsourcing to coordination. In J. Salerno, S. J. Yang, D. Nau, and S.-K. Chai, editors, *The 4th Internatinoal Conference on Social Computing, Behavioral Modeling, and Prediction (SBP)*, volume 6589 of *Lecture Notes in Computer Science*, pages 197–204, College Park, MD, March 2011. Springer. DOI: 10.1007/978-3-642-19656-0. 3

[22] M. Gary and D. Johnson. *Computers and Intractability: A Guide to the Theory of NP-completeness.* WH Freeman and Company, New York, 1979. 39

[23] B. Glavic and K. R. Dittrich. Data Provenance: A Categorization of Existing Approaches. In *BTW '07: 12. GI-Fachtagung für Datenbanksysteme in Business, Technologie und Web*, pages 227–241. Verlagshaus Mainz, Aachen, March 2007. 13

[24] J. Golbeck. Combining provenance with trust in social networks for semantic web content filtering. In L. Moreau and I. Foster, editors, *Provenance and Annotation of Data, International Provenance and Annotation Workshop, IPAW 2006, Chicago, IL, USA, May 2006, Revised Selected Papers*, volume 4145, pages 101–108. Springer-Verlag Berlin Heidelberg, May 2006. DOI: 10.1007/11890850. 12

[25] J. Goldenberg, B. Libai, and E. Muller. Talk of the Network: A Complex Systems Look at the Underlying Process of Word-of-Mouth. *Marketing letters*, 12(3):211–223, 2001. DOI: 10.1023/A:1011122126881. 31, 33

[26] J. Goldenberg, B. Libai, and E. Muller. Using Complex Systems Analysis to Advance Marketing Theory Development: Modeling Heterogeneity Effects on New Product Growth through Stochastic Cellular Automata. *Academy of Marketing Science Review*, 9(3):1–18, 2001. 31, 33

[27] R. Goolsby. On cybersecurity, crowdsourcing, and social cyber-attack. *Commons Lab Policy Memo Series*, 1, 2013. 9

[28] P. Gundecha, G. Barbier, and H. Liu. Exploiting vulnerability to secure user privacy on social networking site. In *The 17th ACM SIGKDD Conference on Knowledge Discovery and Data Mining*, San Diego, CA, August 20-24 2011. to appear. DOI: 10.1145/2020408.2020489. 5, 23

[29] P. Gundecha, Z. Feng, and H. Liu. Seeking provenance of information in social media. Technical report, Arizona State University, 2013. 36, 42

[30] P. Gundecha and H. Liu. Mining Social Media: A Brief Introduction. *Tutorials in Operations Research*, 2012. 1, 4

[31] P. Gundecha, S. Ranganath, Z. Feng, and H. Liu. A Tool for Collecting Provenance Data in Social Media, In *Proceedings of the 19th ACM SIGKDD Demonstration*, 2013. 30, 61

[32] P. Gundecha, Z. Feng, and H. Liu. Recovering Information Recipients in Social Media via Provenance, In *The IEEE/ACM International Conference on Advances in Social Networks Analysis and Mining*, Niagara Falls, Canada, August 25–28, to appear, 2013. 31, 40

[33] O. Hartig. Provenance information in the web of data. In *Proceedings of the Linked Data on the Web LDOW Workshop at WWW*, volume 39, pages 1–9, 2009. 12

[34] R. Hasan, R. Sion, and M. Winslett. Preventing history forgery with secure provenance. *Trans. Storage*, 5(4):1–43, 2009. DOI: 10.1145/1629080.1629082. 23

[35] X. Hu and H. Liu. Social status and role analysis of palin's email network. In *Proceedings of the 21st international conference companion on World Wide Web*, WWW '12 Companion, pages 531–532, New York, NY, USA, 2012. ACM. DOI: 10.1145/2187980.2188112. 30

[36] B. Huberman, D. Romero, and F. Wu. Social Networks that Matter: Twitter under the Microscope. *Available at SSRN 1313405*, 2008. DOI: 10.2139/ssrn.1313405. 42

[37] M. Hurst. Farewell to BlogPulse. Blog, January 2012. Referenced Feb 20, 2012. 1

[38] J. Jonas. Threat and fraud intelligence, las vegas style. *IEEE Security & Privacy*, 4(6):28 –34, nov.-dec. 2006. DOI: 10.1109/MSP.2006.169. 28

[39] A. M. Kaplan and M. Haenlein. Users of the world, unite! the challenges and opportunities of social media. *Business Horizons*, 53(1):59–68, Jan 2009. DOI: 10.1016/j.bushor.2009.09.003. 1

[40] D. Kempe, J. Kleinberg, and É. Tardos. Maximizing the Spread of Influence through a Social Network. In *ACM SIGKDD*, 2003. DOI: 10.1145/956750.956769. 31, 33, 35

[41] G. Kossinets, J. Kleinberg, and D. Watts. The Structure of Information Pathways in a Social Communication Network. In *Proceeding of the 14th ACM SIGKDD international conference on Knowledge discovery and data mining*, pages 435–443, 2008. DOI: 10.1145/1401890.1401945. 42

[42] H. Kwak, C. Lee, H. Park, and S. Moon. What is twitter, a social network or a news media? In *Proceedings of the 19th International World Wide Web Conference*, pages 591–600, 2010. DOI: 10.1145/1772690.1772751. 55

[43] T. Lappas, E. Terzi, D. Gunopulos, and H. Mannila. Finding Effectors in Social Networks. In *Proceedings of the 16th ACM SIGKDD*, pages 1059–1068, 2010. DOI: 10.1145/1835804.1835937. 32, 35, 38, 39, 48, 56

[44] J. Leskovec, L. Backstrom, and J. Kleinberg. Meme-tracking and the dynamics of the news cycle. In *Proceedings of the 15th ACM SIGKDD international conference on Knowledge discovery and data mining*, pages 497–506. ACM, 2009. DOI: 10.1145/1557019.1557077. 58

[45] J. McAuley and J. Leskovec. Learning to discover social circles in ego networks. In *Advances in Neural Information Processing Systems 25*, pages 548–556, 2012. 10

[46] M. Miller, C. Sathi, D. Wiesenthal, J. Leskovec, and C. Potts. Sentiment Flow Through Hyperlink Networks. In *Fifth International AAAI Conference on Weblogs and Social Media*, 2011. 45

[47] L. Moreau. The foundations for provenance on the web. *Foundations and Trends in Web Science*, 2:99–241, 2009. DOI: 10.1561/1800000010. 23

[48] N. Naveed, T. Gottron, J. Kunegis, and A. C. Alhadi. Bad news travel fast: A content-based analysis of interestingness on twitter. In *Proceedings of the 3rd International Conference on Web Science*, 2011. 51, 55

[49] M. Newman. Spread of epidemic disease on networks. *Physical Review E*, 66(1):016128, 2002. DOI: 10.1103/PhysRevE.66.016128. 31, 35

[50] M. Newman. *Networks: An Introduction*. Oxford Univ Pr, 2010. 31

[51] T. O'Reilly and S. Milstein. *The Twitter Book*. O'Reilly Media, Inc., Sebastopol, CA, 2009. 1

[52] S. Petrović, M. Osborne, and V. Lavrenko. Rt to win! predicting message propagation in twitter. In *Proceedings of the Fifth International AAAI Conference on Weblogs and Social Media*, pages 586–589, 2011. 51

[53] B. Prakash, J. Vrekeen, and C. Faloutsos. Spotting Culprits in Epidemics: How many and Which ones? In *Proceedings of the 12th IEEE ICDM*, 2012. DOI: 10.1109/ICDM.2012.136. 32, 36, 40, 48, 56

[54] C. Rovzar. Here's how the rumor that John Roberts is retiring may have gotten started. New York Magazine, March 2010. Accessed, March 4, 2010. 10

[55] S. Rozsnyai, A. Slominski, and Y. Doganata. Large-scale distributed storage system for business provenance. In *2011 IEEE International Conference on Cloud Computing (CLOUD)*, pages 516 –524, july 2011. DOI: 10.1109/CLOUD.2011.28. 13

[56] D. Shah and T. Zaman. Rumors in a Network: Who's the Culprit? *Information Theory, IEEE Transactions on*, 57(8):5163–5181, 2011. DOI: 10.1109/TIT.2011.2158885. 32, 35, 36, 38, 48, 56

[57] A. Shahid. Shirley Sherrod, ex-usda worker: White house forced me to resign over fabricated racial controversy. *New York Daily News*, July 20, 2010. DOI: 10.1145/1084805.1084812. 8, 10

[58] Y. Simmhan, B. Plale, and D. Gannon. A Survey of Data Provenance in E-science. *ACM Sigmod Record*, 34(3):31–36, 2005. DOI: 10.1145/1084805.1084812. 13

[59] Y. Simmhan, B. Plale, D. Gannon, and S. Marru. Performance evaluation of the karma provenance framework for scientific workflows. In L. Moreau and I. Foster, editors, *Provenance and Annotation of Data*, volume 4145 of *Lecture Notes in Computer Science*, pages 222–236. Springer Berlin / Heidelberg, 2006. DOI: 10.1007/11890850. 22

[60] Y. L. Simmhan, B. Plale, and D. Gannon. A survey of data provenance techniques. Technical Report IUB-CS-TR618, Computer Science Department, Indiana University, Bloomington, IN 47405, 2005. 13

[61] B. Suh, L. Hong, P. Pirolli, and E. H. Chi. Want to be retweeted? large scale analytics on factors impacting retweet in twitter network. In *Proceedings of the 2010 IEEE Second International Conference on Social Computing*, pages 177–184, 2010. DOI: 10.1109/SocialCom.2010.33. 51, 55

[62] L. Tang and H. Liu. *Community Detection and Mining in Social Media*, volume 2. Morgan & Claypool Publishers, 2010. DOI: 10.2200/S00298ED1V01Y201009DMK003. 4, 31, 42

[63] X. Wang, H. Liu, P. Zhang, and B. Li. Identifying information spreaders in twitter follower networks. Technical Report TR-12-001, School of Computing, Informatics, and Decision Systems Engineering, Arizona State University, Tempe, AZ 85287, USA, 2012. 14, 49

[64] S. Wasserman and K. Faust. *Social Network Analysis: Methods and Applications*, volume 8. Cambridge university press, 1994. DOI: 10.1017/CBO9780511815478. 42

[65] M. Wesch. An Anthropological Introduction to YouTube. Presentation at the Library of Congress/Electronic, June 2008. Contributors include and The Digital Ethnography Working Group at Kansas State University; Accessed on 22 Mar 2010. 8

[66] R. Zafarani and H. Liu. Connecting corresponding identities across communities. In *the 3rd International Conference on Weblogs and Social Media (ICWSM)*, 2009. 30, 58

[67] T. R. Zaman, R. Herbrich, J. V. Gael, and D. Stern. Predicting information spreading in twitter. In *Computational Social Science and the Wisdom of Crowds Workshop*, 2010. 51

[68] A. Zelikovsky. A Series of Approximation Algorithms for the Acyclic Directed Steiner Tree Problem. *Algorithmica*, 18(1):99–110, 1997. DOI: 10.1007/BF02523690. 42

[69] J. Zhang and H. V. Jagadish. Lost source provenance. In *EDBT '10: Proceedings of the 13th International Conference on Extending Database Technology*, pages 311–322, New York, NY, USA, 2010. ACM. DOI: 10.1145/1739041.1739080. 13, 22

Author's Biography

GEOFFREY BARBIER

Geoffrey Barbier earned his Ph.D. in Computer Science at Arizona State University in Dec 2011. He was a student in the Data Mining and Machine Learning (DMML) Laboratory. He is a 2009 Science, Math, and Research for Transformation (SMART) scholarship recipient. His research interests include social computing, applying data mining and machine learning to social media data, and leveraging crowdsourced data to improve Humanitarian Aid and Disaster Relief (HADR) efforts. He earned a bachelor's degree in computer science at Brigham Young University, Provo, Utah. Geoff also completed a master's degree in business administration through Webster University. He is currently employed as a senior computer scientist at the Air Force Research Laboratory. Any views expressed in this work are the author's personal views and not necessarily those of the Department of Defense or Federal Government.

ZHUO FENG

Zhuo Feng is a post-doc researcher at the Data Mining and Machine Learning (DMML) Laboratory of Computer Science and Engineering, Arizona State University (ASU). He obtained his Ph.D. in the Department of Systems and Industrial Engineering at the University of Arizona. His research interests include social computing, data mining, optimization, and machine learning. His recent research focuses on information diffusion and provenance issues in social media.

PRITAM GUNDECHA

Pritam Gundecha is a computer science Ph.D. student at Arizona State University (ASU). He also works as a graduate research assistant at the Data Mining and Machine Learning (DMML) Laboratory of Computer Science and Engineering at ASU. His research interests include social computing, data mining, and machine learning. His research focuses on security, privacy, and trust issues in social media. He earned a master's degree in computer science at ASU in 2010. He was interviewed by ReadWriteWeb, New Scientist, and Toronto Star for his recent work that was mentioned at more than a dozen media sites. For contact information and links to recent publications, please visit http://www.public.asu.edu/~pgundech/.

HUAN LIU

Huan Liu is a professor of Computer Science and Engineering at Arizona State University. He obtained his Ph.D. in Computer Science at the University of Southern California and B.Eng. in Computer Science and Electrical Engineering at Shanghai JiaoTong University. Before he joined ASU, he worked at Telecom Australia Research Labs and was on the faculty at the National University of Singapore. He was recognized for excellence in teaching and research in Computer Science and Engineering at Arizona State University. His research interests are in data mining, machine learning, social computing, and artificial intelligence, investigating problems that arise in real-world, data-intensive applications with high-dimensional data of disparate forms, such as social media. His well-cited publications include books, book chapters, and encyclopedia entries as well as conference and journal papers. He serves on journal editorial boards and numerous conference program committees and is a founding organizer of the International Conference Series on Social Computing, Behavioral-Cultural Modeling, and Prediction (`http://sbp.asu.edu/`). He is an IEEE Fellow and an ACM Distinguished Scientist. Updated information can be found at `http://www.public.asu.edu/~huanliu`.

Printed in the United States
by Baker & Taylor Publisher Services